Charles Dixon

The Migration of Birds

An Attempt to Reduce Abian Season-Flight to Law

Charles Dixon

The Migration of Birds
An Attempt to Reduce Abian Season-Flight to Law

ISBN/EAN: 9783337253479

Printed in Europe, USA, Canada, Australia, Japan

Cover: Foto ©berggeist007 / pixelio.de

More available books at **www.hansebooks.com**

THE
MIGRATION OF BIRDS

AN ATTEMPT

To Reduce Avian Season-Flight to Law

By CHARLES DIXON

AUTHOR OF

'RURAL BIRD-LIFE,' 'EVOLUTION WITHOUT NATURAL SELECTION,' 'OUR RARER
BIRDS,' 'ANNALS OF BIRD-LIFE,' 'STRAY FEATHERS FROM MANY BIRDS,'
'IDLE HOURS WITH NATURE,' 'THE BIRDS OF OUR RAMBLES,'
'THE GAME BIRDS AND WILD FOWL OF THE BRITISH
ISLANDS,' ETC.

PART AUTHOR OF 'A HISTORY OF BRITISH BIRDS.'

". . . . Wild birds that change
Their season in the night, and wail their way
From cloud to cloud. . . . ."

LONDON: CHAPMAN AND HALL, Ld.
1892

RICHARD CLAY & SONS, LIMITED,
LONDON & BUNGAY.

PREFACE.

THERE is no branch of Ornithology more popular
than that which treats of the Migration of Birds.
To the genuine lover of birds there is no more
fascinating pursuit than to watch the comings and
the goings of his favourites ; to the more scientific
ornithologist Migration is not only an intensely
interesting proceeding in itself, but a function
fraught with importance in the History of Avian
Life. In many instances it is an indicating medium
of affinities, an explanation of various apparent
anomalies in geographical distribution, and unquestionably an evidence of those vast physical
changes which have been one of the dominating
features of our planet's history in past ages.

Notwithstanding the immense popularity and
importance of Migration, strange as it may seem,
no work has hitherto been devoted expressly to
its discussion. A very large amount of material

bearing on Migration has been recorded, and an
equally large amount of observations has been
made on this grand Avian Movement, but hitherto,
so far as I am aware, no naturalist has endeavoured
to grapple with the entire Phenomenon, or to record
the result of its general study in book form. I
am well aware of Palmén's endeavours, of Gätke's
efforts, but both these distinguished naturalists
have only dwelt upon a portion of the subject. I
am equally cognizant of the researches of Weisse-
mann, Harvie-Brown, Cordeaux, Seebohm, Coues,
and Allen, and a host of others; yet none have
sought to exhaust the subject, even superficially,
or to bring our present knowledge of Migration
within the limits of order, or to reduce it to Law.

In the present volume I have made an attempt
to do this. It embodies the results of twelve years
of diligent general study and research, and of at
least two years' close application and thought, and
will, I earnestly hope, serve at least the humble
purpose of paving the way towards a more important
record. The deep interest attached to the important
function of Avian Hibernation has, I hope, been
revived. It would have been an easy matter to
have doubled the size of the present work had I

elected to clothe my facts more heavily with incident, or to confirm my views and opinions' more emphatically with many additional instances. I trust, however, that the elaborateness and length at which the various points have been treated will at least be sufficient to create interest, to stimulate research, and to illustrate the Phenomenon of Migration sufficiently well. I am deeply indebted to a great number of illustrious workers at the subject, not only for material which they have furnished in their published records, but in various other ways, and all of which I trust has been fully acknowledged. To several valued and distinguished direct correspondents my thanks are also due. In the course of my study of Migration many bye-paths of research have been suggested, which it has only been possible to indicate and not to follow. One thing, however, has impressed me profoundly during my years of investigation. In my opinion the study of Migration foreshadows great discoveries relating to the Origin of Species, and the present and past distribution of Life over the earth's surface.

The whole subject of Migration is so vast, so wide-reaching, and so complicated, that it would be absurd to regard it as exhausted, and the present

volume must be looked upon only as a pioneer. If
I have failed thoroughly to grasp its Philosophy, its
History, and its Purpose, may I at least be allowed
to claim the credit of honestly endeavouring to
rescue its mass of raw and tangled data from chaos,
and to reduce them to some kind of system and
Law. No one is more conscious than the author
himself of the little that has been done, and of the
amount of work that still remains to be accomplished ; but if the present effort forms a basis
for more elaborate study and detailed research, its
end and purpose will have been amply attained.

CHARLES DIXON.

July, 1892.

CONTENTS.

CHAPTER I.

ANCIENT AND MODERN VIEWS ON MIGRATION.

CHAPTER II.

GLACIAL EPOCHS AND WARM POLAR CLIMATES.

CHAPTER III.

THE PHILOSOPHY OF MIGRATION.

CHAPTER IV.

ROUTES OF MIGRATION.

CHAPTER V.

EMIGRATION AND EVOLUTION.

Confusion between Emigration and Migration—Defini-
tion of Terms—Theories of Avian Distribution—
Primary Causes of Emigration—Local Causes of
Emigration—Irruptic Emigration—Irruptions of Sand
Grouse—Irruptions of Pastors, Jays, and Goldcrests—
Chronic Emigration—Birds extending their Range—
Present Lines of Migration an Indication of Past
Routes of Emigration—Ancient Routes of Emigra-
tion — Emigration of Ouzels and Snipes—Recent
Avian Emigration in the British Islands—Emigrations
of House Sparrow—Emigration resulting in Migra-
tion—Emigration and Evolution—North Polar Emigra-
tion caused by Glacial Epoch—Island Species of
Birds—Islands and Routes of Migration—Influence
on Insular Avifaunas—Avifauna of the Galapagos

CHAPTER VI.

INTERNAL MIGRATIONS AND LOCAL MOVEMENTS.

CHAPTER VII.

NOMADIC MIGRATION.

CHAPTER VIII.

THE PERILS OF MIGRATION.

CHAPTER IX.

THE DESTINATIONS OF THE MIGRANTS.

CHAPTER X.

THE SPRING MIGRATION OF BIRDS.

CHAPTER XI.

THE AUTUMN MIGRATION OF BIRDS.

CHAPTER XII.

MIGRATION IN THE BRITISH ISLANDS.

THE MIGRATION OF BIRDS.

CHAPTER I.

ANCIENT AND MODERN VIEWS ON MIGRATION.

Ancient Writers on Migration—Their Clear Ideas on the
Subject—Migration to the Moon—Old Observers of
Migration—Mystery surrounding Migration—Transforma-
tion—Hibernation—The Theory still believed in—Evidence
for and against—The Theory of Blind Instinct—Its Impos-
sibility—The Simplicity of Migration—Not a Universal
Habit—Instances Confirming this—Results of Lapsed
Migrations—The Law of Lapsed Migrations—Necessity of
Migration in Autumn—Love of Home—Impulse to Migrate
Hereditary—Antiquity of Migration—Its Ancient Origin—
Necessity for seeking the Cause of Migration in Past Ages.

SIX hundred years before the Christian Era, the
arrival and departure of migratory birds had
undoubtedly arrested the attention of mankind, and
the comings and goings of certain species specially
remarkable for their periodical flights then as now,
were the subject of allusion by the Divine writers
of the Holy Scriptures. In simple yet eloquent
language the Migration of birds is thus commented
upon by the prophet Jeremiah : " Yea, the Stork in
the heaven knoweth her appointed times ; and the

B

Turtle and the Crane and the Swallow observe the time of their coming." There can be little doubt that the migration of birds was equally well observed in even more remote ages, and long before the economy of birds became a subject for scientific investigation, or Ornithology was even in its earliest infancy. For we are informed that the Persians and the Arabs were in the habit of compiling portions of their calendars from the times of arrival and departure of migratory birds, and that the date of their appearance was marked by certain festivals held in honour of the return of a warmer season which these feathered wanderers unerringly proclaimed.

It is rather a remarkable fact that these earliest observers of migration are in no way responsible for the mystery, superstition, and wild incredible theories that have been interwoven with the periodical movements of birds or propounded in explanation of the phenomenon. With the gradual growth of Ornithology as a science the wildest opinions have been expressed, and the most absurd theories put forward concerning migration. From the very earliest times the migration of birds has been a subject endowed with no ordinary degree of fascination for even the most casual observer of animal life. Birds came and went at their appointed seasons, but their destination was cloaked in the impenetrable mystery that surrounded so great a part of the earth's surface in those early days of modern science. Remarkable

as it may seem to naturalists nowadays, it was
gravely asserted, not a century and a half ago,
that the moon was supposed to be the destination
of migratory birds! More than 300 years ago,
we hear of Belon watching with interest the great
flights of various Raptorial birds migrating to
and from their winter quarters. About a century
and a half ago, dear old Gilbert White, Thomas
Pennant, and Daines Barrington, the fathers of
British Field Natural History, were busily engaged
in watching the movements of migratory birds,
corresponding with each other on the subject, and
deeply engrossed with the fascinating pursuit.
Scores of other observers of less eminence in the
scientific world, but none the less earnest in their
endeavours, were studying the periodical flights of
our commoner birds of passage. Continental field
naturalists before and contemporaneous with the
great Linnæus, were similarly employed; and
although their researches were not very systematic
or elaborate, we have ample evidence to show that
the subject was one of no ordinary interest to them.
From the distant days of Gilbert White, who in his
peaceful Hampshire home noted with loving care
the comings and the goings of our feathered hosts,
and never neglected an opportunity of pursuing his
search after migration knowledge, down to our
own, covers the greater part of 200 years, and
now as then the subject has lost no portion of
its charm. In spite, however, of the great pains
taken by these early observers, the philosophy of

migration still remained but little understood, and it may be safely asserted that only within the past twenty years or so has any real light been thrown upon the subject.

Probably no other portion of the economy of birds has been surrounded with so much mystery as that which embraces their periodical journeys to and fro. The various migratory birds were wont to disappear at stated seasons, and to re-appear at others. Such movements were too palpable to be denied. The Cuckoo appeared in spring, and just as surely vanished from the woods and fields in early autumn; so likewise did the Swallows and the Swifts, and the various small birds that swarm in our islands with each recurring summer. Before the geographical dis-tribution of birds had advanced to the dignity of a science, before the uttermost parts of the earth were scrutinized by competent observers, and the avifaunas of wide tracts of country became known through the indomitable pluck and per-severance of collectors, the destinations of these birds of passage were very imperfectly known; and little surprise can be felt that men sought to explain the disappearance of these birds in another way. In other words the Cuckoo, the Swift, and the various species of Swallows were said not to leave this country at all, but actually to spend their winters with us, either, as in the case of the Cuckoo by becoming transformed into a Hawk, or, as in the case of the Swifts and Swallows by sinking into

a state of torpidity for the cold season. We may
at once dismiss the theory of seasonal transmutation as being utterly false and worthless, although
it is soberly believed in even at the present day by
many country folks. It probably originated in the
close resemblance of the Cuckoo to the various
species of smaller Hawk, and viewed in this light is
not so very outrageous after all.

The presumptive habit of Hibernation, however,
cannot be so curtly dismissed. For upwards of
250 years the hibernation of birds has more or less
excited the curiosity of man, and amongst its most
ardent supporters may be found the names of men,
neither knaves nor fools, but eminent for their
scientific knowledge, or renowned for their labours
in the field of Ornithology. Animals both lower
and higher in the organic scale than birds are
known to hibernate, or to pass the cold winters of
the northern hemisphere in a state of torpidity.
Bats, dormice, and various other mammals are in
the habit of regularly hibernating in some snug
retreat during the winter months, in many cases
making elaborate provision for their trance ; various
reptiles, batrachians, and insects hide themselves
away at stated periods and sink into lethargic
slumber. In the hibernation of birds therefore we
have nothing absolutely impossible, either from a
physical or a physiological point of view; and it is
by no means improbable that the habit may have
been a prevailing one in northern regions during
much colder periods than we now experience, and

before migration began. Nay, more, it may have been an inherited faculty from semi-reptilian ancestors now almost become obsolete. The supreme disdain of modern naturalists, almost without exception, for avian hibernation, and the contemptuous way in which they pass the matter by as too utterly absurd for serious notice or refutation, is much to be regretted. No careful student, anxious for truth at any cost, can afford entirely to ignore the mass of evidence accumulated by our forefathers in support of the hibernation of birds; and it seems to the present writer most unwise and most unscientific to consign all this confirmatory material, much of it of the most positive and conclusive kind, to the limbo which contains such absolutely proved fables as the tree-grown Bernicle Goose, the Phœnix, and the Griffon. To many naturalists it may therefore seem undignified to discuss such a subject in sober earnest at all in the present year of grace, and in the highest degree unorthodox to seek to resuscitate a theory so seemingly marvellous, so wildly improbable, as the seasonal torpidity, or periodical decline of vital activity, in certain members of the class Aves!

It is difficult to say how long ago hibernation was first attributed to birds, but it is probably a very ancient assertion, inasmuch that it is alluded to by Aristotle as an opinion prevailing in some countries, and one Olaus Magnus, a northern naturalist, strongly asserted the subaqueous hibernation of Swallows; there is evidence that this

peculiar habit was believed in hundreds of years
ago in Scandinavia, Germany, and elsewhere. Such
accomplished men as Linnæus, Buffon, and Cuvier,
were supporters of the theory. The whole subject
has been repeatedly discussed before such illustrious
bodies as the Royal Society of England and the
French Academy of Sciences, finding a place in the
archives of each. Positive statements have also
been recorded in our own *Philosophical Transactions*, as well as in the Memoirs of the American
Academy of Arts and Sciences, and elsewhere.
Again, it is not a little remarkable that, speaking so
far as the British Islands are concerned, of some
twenty-five common summer migrants, the Swallows
(collectively), and Swift (*Cypselus apus*), are almost
the only species that have been said to hibernate.
Now hibernation was attributed to birds in our
islands so early in the annals of Ornithology that
the probability is these species were the only ones
whose disappearance in autumn was remarked at all.
No other birds that visit us in spring are so
noticeable as these birds : they spend their lives
in the open, coursing about the air, and haunting
the very dwelling-places and cities of men. Small
wonder then that their appearance in spring and
their disappearance in autumn so regularly and so
suddenly fixed the attention of observers, and
excited their curiosity. The migration of birds
was little understood in those days, and double
flights of five thousand miles between Africa and
England were yet undreamed of.

Before proceeding further with this interesting subject, it will now be necessary to say a few words on what avian hibernation really consists of. Hibernation, according to the testimony of observers, may aptly be divided into two kinds. The first division we may designate as subaqueous hibernation, in which birds were said to plunge under the surface of water, and to bury themselves in the mud at the bottom. The second division we will call terrestrial hibernation, and is similar to that prevailing amongst bats and various other mammals, in which birds were said to hide themselves away in crevices of rocks, in hollow trees, and such-like warm and sheltered nooks, there to sink into lethargic slumber until the return of a warmer temperature. So far as the actual process of hibernation is concerned, it does not appear to differ in any way from that undergone by bats or other mammals ; the vital functions are partially arrested, animation is suspended, and a death-like trance or stupor, a lethargic sleep, eventually supervenes. Birds that are said by observers to have been discovered in this state, have slowly regained animation upon being subjected to a warmer atmosphere or to any external heat.

It would be a very easy matter to fill scores of pages with what appear to be well-authenticated instances of the hibernation of birds ; but it will be sufficient for our purpose to allude to a few of the most striking and authoritative ones. In 1666, Schefferus records in the *Philosophical Transactions,* that Swallows sink into lakes in autumn, and hibernate in a manner precisely similar to frogs. In 1741, Fermier-Général Witkowski made legal testimony to the effect that two Swallows had been taken from a pond at Didlacken in his presence, in a torpid state; that they eventually regained animation, and after fluttering about, died some three hours after their capture. In 1748, the great Swedish chemist Wallerius, wrote that he had on several occasions seen Swallows clustering on a reed, until they all disappeared beneath the surface. In 1750, Kalm the traveller observed Swallows on the 10th of April, sitting on posts near the sea, with their plumage wet as though they had just emerged from the water. Four years later, J. R. Forster (editor of Kalm's *Travels in North America*) was an eye-witness, so he informs us, to the following. In January 1754, several Swallows were taken from the lake of Lybshau, then covered with ice, one of which he carried home, where it regained its vitality, but died soon afterwards. In 1764, Achard, referring to Swallows on the Rhine, states that they have been found apparently stiff and lifeless in holes in sand-cliffs, ultimately becoming reanimated. This record doubtless refers to the Sand Martin (*Cotyle riparia*), and is interesting. Between the years 1767 and 1780, Gilbert White's letters to Thomas Pennant and Daines Barrington (the latter then a vice-president of the Royal Society of England) contain many references to the hibernation of birds, and clearly show that the old Hampshire naturalist, one of the most practical, painstaking, and reliable observers of Nature that ever lived, believed implicitly in the habit, although, in spite of much careful investigation, no direct proof of its truth was ever obtained by him. Both Pennant and Barrington were also supporters of the hibernation theory. It should, however, be remarked that both White and Pennant were cautious enough to say that the habit was by no means universal with the Swallows and Swift. In White's letter to Pennant, dated November 4th, 1767, the following passage occurs: "I acquiesce entirely in your opinion that, though most of the Swallow kind may migrate, yet that some do stay behind, and hide with us during the winter." And again in his letter to Daines Barrington, dated March 9th, 1772: "From repeated accounts which I meet with, I am more and more induced to believe that many of the Swallow kind do not depart from this island, but lay themselves up in holes and caverns, and do, insect-like and bat-like, come forth at mild times, and then retire again." After his long life of assiduous observation, White sticks to his honest belief in hibernation, and writes only thirteen years before his death: "Summer birds are, this cold and backward spring [1780], unusually late; I have seen but one Swallow yet. This conformity with the weather convinces me more and more that they sleep in the winter."

For the next fifty years, ornithological literature is fairly well sprinkled with notes on the hibernation

of Swallows and other birds, among the most interesting and precise being the reputed discovery on
the 16th November, 1826, of five Barn Swallows
(*Hirundo rustica*), huddled together in a torpid state
on one of the rafters supporting the roof of a cart-
shed near Loch Awe, in Scotland. These birds
were said to remain some time in an apparently
lifeless state, until the warmth of the room into
which they had been carried roused them into
activity. Other instances of Swallows being found
in a state of torpor in a hollow tree at Belleville, in
North America, and of Sand Martins (probably)
in a sand-bank near Stirling, are worthy of passing
mention. Two instances of Corn Crakes (*Crex
pratensis*), reputed to have been discovered in a
state of hibernation, are recorded in the *Edinburgh
Journal* (vol. viii.). The first example was found
in a mud wall at Aikerness, in the Orkneys; in the
second instance, three of these birds were found in
a dung-heap that had long remained undisturbed at
Monaghan, in Ireland. All these birds were in a
state of torpor, but revived under the influence of a
warmer atmosphere. Reputed instances of Humming Birds (TROCHILLIDÆ) becoming torpid when
overtaken by cold, are recorded in the *Philosophical
Magazine* for 1805 (vol. xxii.).

For the next fifty years but little was heard of
the hibernation of birds; its supporters were either
dead and gone, or the theory had been so pilloried
by modern naturalists that, as Dr. Coues forcibly
remarks, it was as much as a virtuous ornithologist's

name was worth for him to so much as whisper hibernation, torpidity, and mud! In 1877, however, with the publication of Palmen's *Ueber die Zugstrassen der Vögel*, the subject was again brought into prominence before British ornithologists by an anonymous reviewer of the work in *Nature*, and once more the theory was subjected by him to the bitterest ridicule, and denounced as folly. Three weeks afterwards, in the same publication (*Nature*), the Duke of Argyll transmits a letter from Sir John McNeill, wherein the latter gentleman explicitly states that he has seen Swallows hibernating in large numbers. I am indebted to the Duke of Argyll for the following interesting details. " I have an anecdote to refer to on the authority of my late brother-in-law, Sir John McNeill, who told me that many years ago when travelling in the East, he had occasion to cross the Tigris or Euphrates, I forget which, and that he saw a large slice of the muddy bank which had been undermined by the current fall away—exposing to view many Swallows which were dormant in holes in the mud, and of which he picked up a number with his own hands. That Swallows do generally migrate when they leave us is an ascertained fact. That they *can* live in a dormant condition, or do *ever* hibernate, is not believed by naturalists, and I doubt if any evidence but their own eyesight would convince them. Sir John McNeill also told me that he once saw a Hoopoe fly into the hole of a tree at Teheran, in Persia,

and that when he went up to the tree the bird was
already so comatose that he caught it by the hand.
This, however, may not have been connected with
any hibernating habit, as a similar circumstance has
twice occurred to myself with two very different
species. One was a Greenfinch, and the other was
a Nuthatch. I saw a small flight of Green Linnets
fly into a bush on the shore at San Remo, and on
going up to the bush, I found one of them apparently asleep or paralyzed, and caught it with
the hand. My son kept it tame for years, and the
bird was quite well. The Nuthatch I saw in a
similar condition, hanging head downwards from
a twig in my garden in London. This bird I also
caught in my hand, and put it into a cage. But
its condition was so transient that it pecked its
way out of the bottom of the cage the same night,
and escaped. The Hoopoe story, therefore, may
have nothing to do with hibernation, although Sir
John did so interpret it. But the story of the
dormant Swallows in the holes of a mud-bank
would seem to be one of true hibernation, and it
is difficult to imagine that his memory could have
been deceived in such a matter. I believe the still-
living Sir Henry Rawlinson, the celebrated Oriental
scholar, was also present at the time. I think it
clear, however, that migration is the almost universal
rule with birds. Hibernation must be a very
exceptional circumstance."

In the *Ornithologisches Centralblatt* for May 1st,
1877, Rohweder certifies to the accuracy of the

observer who had furnished him with information concerning the hibernation of birds therein described. Both these latter incidents were dealt with in a hostile spirit a few weeks later in *Nature*, by Palmen's reviewer. In 1878, the hypothesis of hibernation received by far the most powerful support ever accorded to it in modern times, in the writings of Doctor Coues, an American ornithologist, and one of the most accomplished and industrious naturalists this or any other century has produced. In his first part of the *Birds of the Colorado Valley*, he not only goes very fully into the presumptive habit of hibernation, but he gives the theory all the support of his authority as an ornithologist of the highest eminence. He there boldly states his belief (and I am not aware that he has seen fit to change his opinion), that the American Chimney Swift (*Chætura pelagica*) hibernates in hollow trees, basing his opinion on the fact that this species is not known to winter anywhere out of the United States, nor is it found anywhere in them at that season ; and of its swarming in myriads in hollow trees, and sometimes perishing in those places in such numbers that their remains form solid masses several feet in thickness at the bottom ! It is, however, only fair to say that Dr. Coues, just like Gilbert White, is not by any means a convert to the belief in universal torpidity, but considers that in the majority of cases it is only odd individuals that do so hibernate. As he forcibly remarks, the migration of a million Swallows

into Africa does not prove that some other Swallows
cannot hibernate.   Unfortunately no direct evidence of torpidity has ever come under this
naturalist's observation.

   The following extract is from the *Dundee
Advertiser* of April 1884.   " About four o'clock in
the afternoon of the 13th or 14th of March, the
light-keeper on duty at the Bell Rock Lighthouse
observed a Swallow fluttering in front of the kitchen
window.   After watching it for some little time, he
opened the window and stood aside to see if the
bird would come in.   This in a minute or two it
did, and alighted on the inside of the window-sill.
So exhausted did the little wanderer appear from its
long flight [?] and the buffeting of the weather, that
it allowed itself to be lifted up and put into a cage.
It immediately lay down on the bottom of the cage
and instantly fell fast asleep, remaining in this state
till next morning about eight o'clock (sixteen hours).
So sound were its slumbers that the keepers watching it as it lay could scarcely detect any signs of
life in it, and at times they were almost certain that
it had died.   On awakening at the hour mentioned,
the Swallow was taken out by one of the keepers
and given a drink of water.   It was put back into
the cage again, where it lay in an apparently
dormant condition till 10 a.m., when it was supplied
with more water, under the influence of which, and
the rays of the sun, it became quite lively and
strong.   The kitchen window was now lifted up,
and the bird taken out and laid in the open hand of

Mr. Jack, principal light-keeper. Resting there for a moment, it gave one cheery twitter, and, springing upwards from the outstretched palm, it winged its way in the direction of the land, and was lost sight of in the space of a minute or two."

This bird may probably have spent the winter, dormant, near the lighthouse. Anyway we here have the most trustworthy evidence of a positive kind. If this does not indicate hibernation capabilities amongst certain birds, pray to what else can it be attributed? Birds normally sleep more lightly than any other creatures; the evidently profound slumber of this individual Swallow was highly abnormal, and undoubtedly of a lethargic nature.

There is another, and it seems to me suggestive fact which deserves notice. Incredulous as it may seem, it is nevertheless true, that the winter quarters of the two most northerly ranging Hirundines are practically unknown. The House Martin (*Chelidon urbica*) and the Sand Martin (*Cotyle riparia*) are known to breed in large numbers, in some cases literally to swarm in the Arctic regions, the former reaching latitude $70\frac{1}{2}°$ in Western Europe, and latitude 69° in Siberia; the latter 70° in Western Europe, 67° in Siberia, and in Kamtschatka on the Pacific coast. As is well known, these birds literally swarm in many districts during summer; they are obtrusive species by no means easily overlooked, yet nowhere do we either hear of them on passage, or have they been found in Africa, India, or elsewhere during the cold season in any proportionate numbers. Is it possible that the most northerly birds hibernate and pass the long Arctic winter in torpidity as so many other boreal creatures are known to do? I express an opinion neither one way nor the other, but allude to the facts as suggesting an interesting field of inquiry. It has also been remarked repeatedly that abnormally early individuals of Corn Crakes and Swallows have been noticed in or near districts where unusually late ones were observed during the previous autumn.

The above may be taken as the pith of the evidence in favour of hibernation; it is now only fair that the evidence against it should be briefly discussed. I think we may at once dismiss subaqueous hibernation as applied to birds as a physical impossibility. The species on which this portion of the theory is based could not live for even a few moments under the surface; water is not their element; whilst the sudden arrest of vital activity and the abrupt transition from mercurial energy to torpor is absolutely fatal to the supposition that they should seek such a retreat. We have, however, the Duke of Argyll's interesting experiences of birds suddenly assuming a comatose condition, which may bear directly on hibernation under other conditions. The theory of subaqueous hibernation probably had its origin in the fact that Swallows are particularly fond of frequenting large sheets of water, especially in autumn, at which season vast numbers often resort to reed and osier beds to roost. In skimming to and fro they often drink or bathe, and repeatedly strike the surface with their wings; this may have suggested a disappearance beneath the surface. Of course subaqueous hibernation as applied to amphibious animals is a different thing altogether, and, it need scarcely be remarked, is a well-established fact. Even terrestrial avian hibernation has at least one grave difficulty to contend against, especially when applied to Swallows, the very birds, by the way, to which it has been most widely attributed. These birds undergo their annual change of plumage during the months of February and March, whilst they are in their winter quarters, and it is scarcely conceivable that such a function could be performed during a state of torpor; although, in the case of Swifts, it may be remarked, Gilbert White suggested that they might perhaps retire to rest for a season, and moult in the interval. Swifts, I should say, however, moult twice in the year, and very slowly.

Strange, nay almost incredible as avian hibernation is, however, it must always be remembered that the evidence against it is purely negative; and that although it has not yet been sufficiently established to satisfy the sceptical science of to-day, it has never been refuted. Denials prove nothing, and all we can say is, until more satisfactory modern evidence of its truth is forthcoming, that birds are probably capable of hibernating under exceptional conditions, and may have done so, but this habit is by no means a usual or a universal one, and very likely in most if not all cases has arisen not from choice, but from inability or strong disinclination to migrate at the customary period. Hibernation, so far as we can learn, only applies to a few individuals, and no species of bird has yet been discovered in which the practice is universal, if we except conditionally the Swift (*C. pelagica*), to which allusion has already been made. As for myself, I neither accept nor deny it, having personally seen nothing to refute or confirm it, although fully believing it possible, considering that such an attitude is the most scientific position to assume until the subject has been more fully investigated, even at the risk of being "handled without gloves" by some mud and torpor despising bruiser critic for my heresy!

Little if any less marvellous was that mysterious power ascribed to birds which enabled them to perform their journeys to and fro between continent and continent with such wonderful skill. Even at the present day there are many naturalists who implicitly believe in this miraculous power, and as a popular opinion it still prevails supreme. Birds are said to be endowed with the superhuman faculty of finding their way across the sea to their winter quarters; or setting off as each migration time returns for a flight of many thousands of miles with nothing but their inborn perception or blind instinctive impulse to guide them on the way. In short, birds at the present day are still popularly supposed to migrate by "instinct," knowing not how or why, impelled along a course that never errs or changes, flying from point to point with no more mental effort than a piece of inanimate mechanism, hurrying away as an arrow from a bow, and just as incapable of swerving from their track until their flight is done. Were such a miraculous power really possessed by birds, then of a surety they would be transcendently more endowed with mental attributes than Man, creation's Lord himself! Ascribing such a faculty to these creatures means on the one hand exalting them to a super-human height of intelligence, or on the other hand degrading them to the passionless depths of mere irresponsible automata. Both courses are as unnecessary as they are unscientific and illogical. Migration is a Habit, and like all other habits has had to be acquired; it is, however, by no means a universal habit even amongst the individuals of many species; some migrating, others remaining stationary, according to the exigencies of their surroundings. Nay, more, we might almost say that uniformity is the exception rather than the rule. Were migration an inherited instinct, it is only reasonable to presume that it would be transmitted from parent to offspring in one unbroken and unchanging order of descent.

Let us glance at a few instances. The Robin (*Erithacus rubecula*) is a migrant in all the colder portions of its range, where the winters are sufficiently severe, passing for instance between Scandi-

navia and Africa with great regularity, but in milder
climates, as for instance in the British Islands, it
is stationary. The Sedge Warbler (*Acrocephalus
phragmitis*) is a bird of regular passage, yet many
individuals have suffered their migratory impulses
to lapse, and reside permanently in parts of North
Africa, in Corfu and Crete. The Blackcap (*Sylvia
atricapilla*), although a bird of passage with us, and
notwithstanding the fact that its migrations extend
from about the latitude of the Arctic Circle to within some ten degrees of the Equator, is a resident
in the basin of the Mediterranean. The dainty
little Willow Wren (*Phylloscopus trochilus*), one of
the most familiar of our summer birds of passage,
whose migrations extend from nearly the highest
point of continental land in Europe and Western
Asia to the southernmost parts of Africa, in a fly-
line of nearly 8000 miles, may be taken as one of
the very best examples of migration, and yet in this
species the habit is by no means universal, and
many individuals are resident in North Africa,
Spain, and Sicily. Just as remarkable is the
Chiffchaff (*Phylloscopus rufus*); in most parts of
its range it is a migrant, but many individuals are
stationary in Southern Europe. Indeed, a few odd
birds have been known to forego migration even
in the British Islands and Germany, remaining to
winter in the more sheltered parts of those countries.
Then the beautiful little Goldcrest (*Regulus cristatus*) is only migratory in the colder parts of its
range, being, as we know, resident in our islands;

but its numbers are increased in autumn by individuals from far across the sea.   The Hooded
Crow (*Corvus cornix*) is yet another instance of
modified migration.   In Scotland this bird is
sedentary, elsewhere it is one of the most regular of
migrants; vast numbers, for instance, coming to the
eastern districts of England in autumn and returning in spring.   Even the Magpie (*Pica caudata*)
is a migrant in the northern and colder portions of
its range.   The Reed Bunting (*Emberiza schœniclus*)
and the Common Bunting (*Emberiza miliaria*) are
both examples of species in which many individuals
are of regular migratory habits, whilst others are
practically resident.   Such a sedentary species as
the Yellow Bunting (*Emberiza citrinella*) is in this
country, is a regular migrant from the Arctic
regions, retiring in winter to the South of Europe,
Asia Minor, North-western Persia, and North-
western Turkestan.   The Ring Dove (*Columba
palumbus*) is a regular bird of passage in the
northern portions of its range, passing Heligoland
in large numbers every autumn, and less abundantly
in spring.   Such a well-known migrant as the Corn
Crake (*Crex pratensis*) is almost everywhere a bird
of passage; but in Algeria, Palestine, and Asia
Minor it is sedentary.   The well-known Oystercatcher (*Hæmatopus ostralegus*) is a resident on the
British coasts, but the individuals spending the
summer on the coasts of Arctic Europe and the
Baltic Sea are birds of regular passage to Africa.
In the same manner the Little Ringed Plover

(*Ægialitis minor*) is a bird of passage in the
northern portions of its range, but resident in the
basin of the Mediterranean, the migratory individuals extending as far north as lat. 60° in summer,
and as far south as the Equator in winter.   Even
the Common Sandpiper (*Totanus hypoleucus*) is by
no means uniformly migratory.   Few birds are
more regular in their appearance in spring upon
our upland streams and pools than the dainty
"Summer Snipe," and yet in the basin of the
Mediterranean many individuals have dropped the
habit of migration altogether, and have become
sedentary.   The Redshank (*Totanus calidris*) is
even more irregular in this respect.   In the north
of its area it is migratory, but in the basin of the
Mediterranean it is resident, and yet the winter
passage of many individuals of this species extends
south to the Cape Colony.   Many individuals of
various species of Ducks have also ceased to
migrate, and have become residents in our islands
within comparatively recent times.   Among these
may be mentioned the Tufted Duck (*Fuligula
cristata*) and the Pochard (*Fuligula ferina*).   In
some cases the cessation of migration has led to
physical change, and the segregation of local
southern races.   One instance of this I can here
call to mind is the small Indian race of the Little
Ringed Plover (*Ægialitis jerdoni*).   Another instance is probably presented in the two races
of Ringed Plover (*Ægialitis hiaticula*, and *Æ.
hiaticula major*), the former migratory, the latter

sedentary. Many more instances might be adduced to show that Migration Habit is by no means universal, even amongst the individuals of a species; and even amongst the migratory individuals themselves the length of the journey varies to an astonishing degree, some travelling double or more the distance others travel. From the above facts, we may also propound the law that wherever the breeding area of a species intergrades with its winter range, migration among individuals breeding in the impinging districts has been suffered to lapse.

It is quite unnecessary here to deal with the modification that migration has undergone in past ages, having shown that it is not an unchanging habit, that it is not a transmitted power from parent to offspring in one immutable sequence; we need not therefore stay to inquire into its previous history as regards individual species, or even groups of the early ancestors of our existing avifauna. No doubt species have acquired the habit, and discarded it times without number in the course of their eventful descent from remote ages, according as conditions of life have favoured one or the other end. Birds migrate from necessity, not from choice; in confirmation of which fact I may mention that I do not know of any instance where some or all of the individuals of a species quit their breeding-grounds unless compelled to do so by severity of climate, failure of food, or both. The migrations of some species are shorter than others,

even from high-lands to plains only, yet the movement is a marked and regular one, and does not
differ in kind from those extended flights of thousands of miles which other species habitually undertake.   On the other hand, birds are excessively
attached to their home, to the land of their birthplace and their love.   Witness the fact of so many
species (the Redwing, *Turdus iliacus*, and the Fieldfare, *Turdus pilaris*, are good instances) quitting
their winter quarters in spring to return to their
old home, although their young, so far as we can
determine, could be just as successfully reared in our
islands.   Other interesting instances might be taken
from the Waders, the Ducks, the Gulls, and Terns.

Instinct, again, is described as infallible.   Migrating birds go and come with unerring certainty;
they know their way by inherited impulse, and
never fail to reach their destinations.   But such is
not the case.   Birds blunder like human folk, lose
their way, and perish in uncounted hosts, as we
shall learn anon.   The Homing Pigeon has been
frequently brought forward in support of the
inherited instinctive sense of direction possessed by
birds in general.   But this bird, wonderful as its
performances of speed, distance, and endurance
really are, has to be long and carefully trained for
each successive stage of its prolonged journey
before it can be safely entrusted to undertake it.
During this training it gradually learns the various
landmarks on the road, just as any human traveller
might do ; and if called upon to make the journey,

say in darkness or in fog, invariably declines to
essay the task, or, in the rare event of it doing so,
soon loses its way, its mysterious sense of direction
being of course a myth.   And so it is with migratory
birds in general.

Amongst birds in which the habit of migration
is dominant, the impulse to migrate is unquestionably instinctive, in the sense of being an hereditary
desire transmitted from parent to offspring, which
has become so deeply rooted in the uninterrupted
course of countless ages of passage to and fro, that
in many species nothing but death can eradicate it.
Migratory birds if kept in confinement begin to
grow restless and unsettled as the usual period of
their departure draws nigh ; the same irresistible
desire is reflected in the gathering of the swallows
in autumn ; and the unwonted activity of other
little feathered voyagers among the trees and hedges
may be remarked by any one who takes the trouble
to observe it.   This desire to migrate gradually
becomes an overwhelming desire, before which all
other inclinations bow, and at last the great flight is
commenced.   But here instinct, hereditary desire,
ceases its sway ;   reason, memory, knowledge of
locality, and perception take its place.   The mysterious portion of migration may now be thought
to commence, as these little travellers depart for a
distant land ; but the process is simple in the
extreme.   It is sufficient for the present to say that
all a bird's amazing powers of memory for certain
landmarks and its knowledge of locality are brought

into play—powers, be it remarked, far more acute
than any possessed by human intelligence, as the
great journey progresses; and thus aided the wondrous flight is taken mile after mile along the old
familiar way until the end of that long journey is
reached. We shall return to this particular portion
of the subject, and treat it more elaborately later on.

The antiquity of migration is profound. It is a
habit connecting the present day with the immea-
surable ages of the past, more ancient probably than
any other, save that of reproduction—a habit that
has been possessed from the very earliest infancy of
Avian Life, maybe; handed down unchanged from
a past so remote that the mind of man fails utterly to
grasp its mighty measure. So soon as Birds, beings
capable of *flight*, were evolved from their semi-
reptilian ancestors, and circumstances arose which
caused a change of habitat, the Migration of birds
may be said to have begun, and to have continued
from that remote past more or less intermittingly
until the present day. The key therefore to the
Habit of Migration must be sought in past ages,
and as the subject is necessarily an important and
an extensive one, a chapter must be set apart for its
discussion.

CHAPTER II.

IF we accept the footprints upon the Bunter Sand-
stein of the Triassic System as evidence of the
presence of Birds, then these creatures may probably date their origin from ages so remote as the
beginning of the Mesozoic Period. Dr. Hitchcock
enumerates the footprints of as many as twenty-
three species of " birds " in the Triassic formations
in New England. Some of these are of enormous
dimensions (twenty-two inches in length), and in-

dicate creatures of vast size. Whether these giant
Birds or Bird-reptiles, for we have no positive evidence of their structure, were capable of flight it
is impossible to say. Whether they were sedentary
or migratory in their habits is equally uncertain.
In the Oolitic or Jurassic System, the ages of the
gigantic Reptile and the Pterodactyl, a reptile-like
bat, more definite evidence of the existence of birds
is forthcoming in the famous fossil Archæopteryx.
From this we pass on to the toothed birds of the
Cretaceous System, all of which, however, appear to
have become extinct before the Tertiary Period, in
which occur the fossilized remains of birds closely
allied to existing forms.

We need not concern ourselves with the effects
of Glacial Epochs (even if such occurred, and the
evidence against such phenomena having occurred
earlier than the close of the Palæozoic Period is
very conclusive, as will shortly be seen) previous
to Miocene and Eocene ages ; and what effects the
mighty terrestrial changes may have produced on
then existing life does not bear upon our present
subject. That vast changes in the earth's surface
and in her planetary movements took place in
Miocene and Eocene ages, there is apparently
sufficient geological and astronomical evidence to
suggest ; and in those remote ages Birds—creatures
as we know them to-day—lived and flourished.
We thus see that the Avian order of living beings
has been subjected to many and varied geographical
and astronomical disturbing influences from its

very earliest infancy. The Migration of Birds—
a result of Change, as I hope soon to demonstrate
—therefore dates most probably from Miocene or
even from Eocene ages; for we know that during
those periods vast alterations took place in the
relative level of land and sea, that volcanic agency
was remarkably active in altering the physical
aspect of the northern hemisphere, and that climatic
changes, due to varying phases of the earth's eccentricity and divergence of ocean currents, were considerable. The phenomenon of migration, however,
must then have been very different from what it is
in our day, and therefore, to understand it in its
present aspect, we must pass on to a later period
in the earth's eventful history. That in those
mighty changes the habit of migration had its
origin can scarcely be disputed, but much modification necessarily took place during the occurrence
of the Glacial Epoch succeeding them. This we
may call the Post-Pliocene Glacial Epoch.

That parts of the northern hemisphere have been
subjected to a long and severe period of glaciation
within comparatively recent geological time, is
proved by the fact (among others) that the various
species of Mollusca then living are the same as
those we meet with in our own age; not only so,
but numerous traces of ice movement have been
preserved to us. It is 200,000 years ago since
this Glacial Epoch is computed to have reached
its maximum, and about 80,000 years since it came
to a termination. We need not enter here into

details of the causes of Glacial Epochs, or of the
various theories propounded to explain them. To
such of my readers who may desire to acquaint
themselves further with this portion of the subject,
I would refer to Dr. Croll's *Climate and Time in
their Geological Relations*, and to Wallace's *Island
Life*, in which latter work especially the whole
phenomenon is treated in an original and very
masterly manner. Briefly stated, Glacial Epochs
are caused by the slow and irregular periods of
great eccentricity in the earth's orbit, combined
with the precession of the equinoxes, and initiated
by high land and unusual amount of moisture
(which are favourable to the accumulation and
storage of ice and snow) at the glaciated pole. It
appears pretty certain that periods of high eccentricity cannot produce glaciation unless the land
area of the pole with its winter in aphelion is
favourably adapted, and the warm ocean currents
have been diverted from the polar area. All the
geological evidence hitherto collected is against
the supposition that any one or more Glacial
Epochs occurred during the Secondary and Tertiary Periods. As Wallace clearly demonstrates,
this vast expanse of time seems to have been characterized by a uniform warm or temperate climate,
admitting of a luxuriant vegetable growth up to
the highest latitudes at present reached by man.
The evidence points unquestionably to the fact
that the Post-Pliocene Glacial Epoch was an exceptional phenomenon—no such vast and terrible

glaciation having ever previously occurred in north polar regions within geological time, although this view is not shared by many scientists, including Dr. Croll, who persists in his theory of intermittent or alternate glaciation following phases of high earth eccentricity. As Professor Nordenskjöld remarks (*Geological Magazine*, 1875)—" An examination of the geognostic condition, and an investigation of the fossil flora and fauna of the polar lands, show no sign of a glacial era having existed in those parts before the termination of the Miocene Period, [therefore] we are fully justified in rejecting, on the evidence of actual observation, the hypothesis founded on purely theoretical speculations, which assume the many times repeated alternation of warm and glacial climates between the present time and the earliest geological ages."

The Post-Pliocene Glacial Epoch seems to have been primarily brought about as follows. During the latter part of the Secondary Period a considerable elevation of land occurred in the Arctic regions, apparently the climax of a slow compact development of continental land, which not only barred several of the water passages to the north, but probably checked the flow of various warm ocean currents to this polar area, simultaneously with or just previous to a long phase of high eccentricity of the earth's orbit. From this evidence we may justly infer that birds have had only one experience of a Glacial Epoch, for the one preceding it was probably, if at all, during the Permian Period in

ages before birds were evolved. This Glacial
Epoch, however, did not consist of one gradual
change from a temperate to an ice climate which,
having reached its maximum, again passed slowly
away, but the earlier and later stages of the grand
phenomenon were broken up into several alternative
periods, during which warm and cold climates prevailed respectively in the area affected every 10,500
years in the precession of the equinoxes during this
phase of the earth's high eccentricity. From the
commencement of this Glacial Epoch, the Migration
of birds, as we see it at the present time, was probably initiated. That these warm and cold Glacial
Periods actually took place, we have abundant
palæontological evidence, even in the British Islands.
The fossilized remains of Hippopotami (*Rhinoceros
hemitæchus*) and Elephants (*Elephas antiquus*) testify
to the uniform temperate climate with little or no
winter, just as the remains of Mammoths and Rein-
deer are indicative of a cold Arctic climate, with a
short hot summer and a long severe winter succeeding
it. There seems every inducement also to presume
that during the Glacial Epoch the ice, once formed,
never left the immediate regions of the Pole even
during the several warm inter-glacial periods, which
had such vast influence upon the climate of the
sub-arctic zone.

From these remarks it may be gathered that we
do not require even the occurrence of one Glacial
Epoch to account for the Migration of birds. That
a habit so deeply rooted, so universal, and so vital

to the well-being of such a vast number of species
must have had its origin in a very remote past, the
result of mighty climatic change and physical disturbance, seems not only absolutely certain, but the
only rational explanation of the phenomenon. Such a
cause amply sufficient in every respect is to be found
in varying phases of Earth's orbital eccentricity in
combination with the precession of the equinoxes—
the grandest cycle of discovered Time, which in
wondrous course entirely reverses the seasons of
either hemisphere as the earth's polar axis describes
a complete circle in the heavens. That these
majestic phenomena are in any conceivable way
connected with the migratory movements of birds
seems utterly impossible; but in them the habit has
its root; and the simple season-flight of a Cuckoo
or a Nightingale to and fro between the shores of
Africa and England is inseparably and directly connected with the erratic movement of a planet in its
orbit; nay, with the constitution of a universe!

Of course in the utter absence of any data to fix
the northern range of migratory birds in remote
ages, it is impossible to give detailed explanations of
the cause of the habit in each species, or even in
many groups. Many local influences may have
been at work; the gradual extension of range in
various directions from a centre of dispersal owing
to the rapid multiplication of a species may have
driven many individuals into more northern regions
where only a summer residence was possible; or
even caused longitudinal movements necessitating

the acquirement of a migratory habit from more local causes. It will thus be seen that it is quite impossible to explain the origin of Migration in every species, or in an individual degree, but only to suggest its cause in the broadest sense. With Polar groups it is comparatively easy, as, for instance, in the CHARADRIIDÆ. But in all, broadly speaking, the habit had its origin either in direct climatic change, or in the exposure to a different and more severe climate due to Emigration. Again, some species must have acquired the habit sooner than others, just as we know at the present time the migratory flight of some birds is earlier or later than that of others, according to specific or even individual requirements. From this we must infer that Migration did not have a simultaneous origin ; some species probably remained stationary ages after others had acquired the habit, gradually acquiring it as the causes became more intensified. Broadly speaking, it is birds that subsist on an insectivorous or an animal diet that are the greatest and most regular migrants, and they would be the first to acquire the habit. But even from the great variety of their food many would be compelled to adopt migratory habits before the others. Birds subsisting on a vegetable diet would persist in their sedentary habits the longest, and remain resident so long as their food was not buried in the snow. Many ages, however, must have elapsed between the period when the winters became too severe for insect life to remain active, and the ages when

animal food might be picked up on the Arctic
coasts; or again, the ages when winter had become
so severe that all the land was buried deep in snow,
and the ocean itself sealed with ice. These varying
details, however, do not prevent us from going back
to a remote past, when probably the birds inhabiting
temperate and even northern regions were sedentary.
That this was probably the case in Eocene times is
reflected in the fact of the wondrous wealth of vegetation, indicating a corresponding high degree of
temperature then prevailing, when palms clothed
the English plains, when crocodiles basked in the
rivers, and a glorious sub-tropical climate reigned
supreme. Or even more recently in Miocene ages,
when chestnuts and magnolias, walnuts and vines
flourished in Greenland, and the swamp-cypress and
the water-lily grew on now desolate and ice-doomed
Spitzbergen—when the Lower Miocene climate of
Switzerland, for instance, resembled that of North
Africa to-day. To understand the subject clearly,
it will now be necessary briefly to review the changes
which have undoubtedly occurred during this
remote past of Tertiary time, and of which we
have abundant geological and palæontological evidence, and which are still further confirmed by
astronomical calculation.

From Dr. Croll's published tables (*Philosophical
Magazine*, xxxvi. 1868), showing the amount of
the Earth's eccentricity of orbit for the past three
million years, it appears that a long-continued
period of exceptionally high eccentricity occurred

little less than 2,000,000 years ago, and another about 850,000 years ago, which probably correspond in time with the Eocene and Miocene Periods respectively. Owing to the physical aspect of the great masses of continental land during these ages being such as to admit not one but several warm ocean currents into the Polar area, the differences of climate produced by precession and these great changes of eccentricity would not be very marked. As Wallace remarks: "the summers would be at one period almost tropical, at the other of a more mild and uniform temperate character; while the winters would be at one time somewhat longer and colder, but never probably more severe than they are now in the West of Scotland." Gradually through the Pliocene Period the eccentricity of the earth's orbit again became exceptionally high, and finally culminated in the Post-Pliocene Glacial Epoch which was brought about by this and other causes, already noted. It will thus be seen that although the climate never became glacial during the Tertiary Period, great changes took place, perhaps not sufficiently severe to lead to a very decided Migration of birds, but enough to initiate the habit in a comparatively small degree.

That the Glacial Epoch had an incomparably deeper and more lasting effect upon the movements of birds than any other climatic change during Avian history, is unquestionable. As this grand phenomenon progressed, rendering the climate of the Polar and temperate zones more and more rigorous, the

Migrations of birds became increasingly pronounced, and its culminating point was reached when the North Polar world became covered with a vast icy mantle, and all living things were either killed or banished to more southern latitudes. During the initial stages of the Glacial Epoch, when eccentricity was not so high as at a later phase, alternations of warm and cold periods were caused by winter occurring in perihelion or aphelion. But as long as eccentricity remained high and at its climax, precession had little effect upon the glaciated Pole, and only a partial amelioration of climate took place during the inter-glacial periods in more southern regions. When eccentricity was considerably diminished precession had again a more marked effect upon the climate, and a warmer temperature prevailed than is now the case. Since the passing away of the Glacial Epoch the eccentricity of the earth's orbit rapidly decreased, and for the past 60,000 years it has remained uniformly low, with this inevitable consequence, that the changes of climate produced by precession have been correspondingly slight, and the climate of the north temperate zone has remained in an exceptional state of stability. One inevitable result of the Glacial Epoch was to place Migration on a very different basis from what it had occupied before. All the Præ-Pliocene order of things had passed away; continents had become more closely knit together; seas had vanished; archipelagoes disappeared; warm currents had been diverted and checked; Polar land elevated.

Now I think from the above evidence we may safely come to the conclusion that vast changes of climate have taken place at irregular intervals, not only throughout Tertiary time, but during the entire progress of the Glacial Epoch in the northern hemisphere. Warm tropical climates have prevailed; cold climates have succeeded them; the Polar area has been peopled with a rich fauna and flora (birds of course included); and just as surely has it been devastated and become the habitude of icy desolation, entailing the utter banishment of life. We have thus seen that the Glacial Epoch was not a primary cause of Migration; rather has it been a colossal agent of emigration and of banishment; for many of the avian forms then driven from their Polar homes never returned, but were the founders of innumerable colonies in more southern latitudes. In this banishment many new forms had their origin through Variation, preserved by Isolation and Natural Selection.

It now becomes necessary to show what effect local glaciation may have had on the vertical Migration of birds. There can be little doubt that during the Tertiary Period many birds although resident in the Northern Hemisphere retired from the lowlands to the mountains to breed, or were even residents on such lofty ranges as occurred within this area. Now it seems probable that during phases of high eccentricity, in Miocene ages, for instance, much local glaciation occurred in the Alps, in Scandinavia, the Pyrenees, the Caucasus, the Himalayas, and elsewhere, wherever great extents of high land offered favourable conditions to its initiation. From these localities all living things were driven down the hillsides by the advancing glaciers; although on the lowlands and in the Polar area, where no vast masses of elevated land occurred, and where other geographical conditions were favourable, such a phase of high eccentricity produced little change, and these areas remained in the enjoyment of an uninterrupted climate of a warm character. This local glaciation would have the effect of slowly compelling resident birds to adopt migratory habits, or cause considerable emigration amongst them ; and it is conceivable how birds that visited the uplands every summer gradually modified their movements in various ways to adapt themselves to the changing aspect of their breeding areas. As these glaciers once more retreated, migratory movement would again be excited in many directions ; and one can imagine how birds after becoming residents on the lower grounds, renewed their seasonal visits, or even emigrated on a large scale to the haunts their ancestors held in past ages. Even during such periods of low eccentricity as prevail at the present day, the influence of precession on the snow-line of mountains would be considerable, and its results on the movements of birds important.

A few words now on the probable future of Migration. That it will undergo many important changes during the next precession of the equinoxes is as certain as that that grand revolution of the seasons has produced changes in the past. At the present time the northern temperate zone with winter in perihelion combined with a low degree of eccentricity is favoured with a comparatively mild climate, and this, in the normal course of things, will be enjoyed for some thousands of years. Then, however, with winter in aphelion, a colder climate will undoubtedly prevail; the winters will become colder, and last longer; the summers become hotter, and shorter in proportion. These changes would so modify the present climate of the British Islands that, as Wallace suggests, perpetual snow will rest on all our highest mountains. The effect of such an inevitable change of climate will be to banish many of the birds from our islands during winter that are at present resident with us, to cause many other migrants to alter the period of their annual journeys, and to lower the winter range of many boreal forms that at the present time winter in latitudes north of us. This descent of northern forms and banishment of species to southern haunts will also entail many movements among species wintering in them, through changed conditions of life due to competition, and consequently a fiercer struggle for existence.

It has been suggested by Mr. Seebohm that the migratory habits of the birds comprising the great Polar family CHARADRIIDÆ were initiated by the want of light during the North Polar winter (or the season analogous to such, for no great change of temperature then occurred to separate that period

so acutely from summer as now), when all was in
perpetual star-lit gloom, even during the prevalence
of mild climatal conditions. But I think the
selection is singularly an unhappy one as an
example of this cause of Migration. It seems to
me that the want of light necessitated the habit of
feeding during darkness—a habit which is notoriously continued to the present day, even by
preference, by many of the Waders and Ducks,
probably the birds last to linger in the Polar basin
at the commencement of the Glacial Epoch. Now
the one great dominating impulse to Migration is
undoubtedly the want of food. Whatever influenced this supply, either directly by destroying it
altogether, or indirectly by rendering its capture a
more difficult or even impossible proceeding, would
have an irresistible tendency to cause migratory
habits to be adopted. Polar darkness would have a
direct influence on insect life in two ways. Either
it would cause these creatures to retreat for the
period of its continuance, or render them partially
nocturnal for the like time. Of the two, the
latter seems to me the most likely. It therefore
follows as a natural consequence that the birds
living on insects would either have to retreat south
beyond the limits of Arctic darkness, to where light
prevailed for a sufficient time each day to enable
them to find sufficient food; or, in the event of
insects becoming nocturnal, to so modify their own
habits as to become nocturnal themselves. That
the Waders adopted the latter course seems proved

by their strong addiction to nocturnal feeding now, notwithstanding the fact that their food (mollusks, crustaceans, worms, and insects) may now as then be obtained at all hours of the twenty-four. Of course during spring and autumn they would be open to the necessity or the inclination perhaps of feeding by daylight as well as by night, whilst in summer their nocturnal habits would be lost in eternal day. Darkness would have little or no effect on seed-eating birds, for their food supply would receive no modification, and be as accessible in the dark as in the light. Besides, these birds then as now did not probably range into such high latitudes as to be subject to the influence of Polar darkness; they doubtless dwelt in lower latitudes where the day in winter would be short, but sufficient for their requirements; and this fact is confirmed in what I believe to be a very startling way by the habit of all seed-eating birds rising late and retiring early— *with the sun, in fact*—even at the present day. A tendency to nocturnal habits is the rule rather than the exception among all northern and insectivorous birds. The water birds are notoriously as much at ease either by day or by night in their quest for food. It is also a most remarkable fact that a very high percentage of Arctic insectivorous Passerine birds show a decided partiality for *feeding at dusk*. Habits like these are neither acquired without cause nor readily relinquished. The Redwing (*Turdus iliacus*), the Fieldfare (*Turdus pilaris*), and the Arctic Bluethroat (*Erithacus suecica*) are all birds breeding at or above the Arctic Circle (and doubtless bred much farther north during milder climatic conditions), and still possess the habit of feeding far into the twilight.   This, I consider, seems to imply that they or their ancestors (for the habit widely prevails throughout the TURDIDÆ) continued to inhabit the northern world during the Polar night, so long as climatal conditions permitted a constant residence in these regions both of themselves and the various creatures on which they fed.   That want of light may have been an initiating cause of Migration is not altogether improbable; but a gradually lowering temperature was undoubtedly the most potent, and as applied to the Waders and the Ducks the only cause.

It has been laid down as an axiom in Ornithology that the birds that go the furthest north to breed go the furthest south to winter; but the vast importance of this fact as bearing on the past history of Migration appears hitherto to have escaped notice.   It is an interesting fact in itself, but to my mind the cause of the habit is immeasurably more interesting, indicating, I would suggest, an Ancient Migration extending almost from Pole to Pole. That the Antarctic continent (a circular mass of land more than twice the area of Australia) during remote ages has from time to time enjoyed periods of climate mild enough to support a fauna and flora as rich as or even much richer than those now peculiar to Arctic lands, seems indisputable. The botanical evidence is certainly in favour of such conditions having prevailed in past ages; there is much evidence that the centre of dispersal of many groups of birds is Antarctic.

Now it is a remarkable and suggestive fact that among the CHARADRIIDÆ the species breeding in what we may well describe as the North Polar Basin have the longest migrations of any known birds. The following half-dozen species are among the most Arctic of birds. First we may instance the Sanderling (*Calidris arenaria*), breeding in the North Polar Basin, and extending its winter migrations southwards to the Malay Archipelago, the Cape Colony, and Patagonia; second, the Turnstone (*Strepsilas interpres*), whose breeding range is circumpolar, and extends into the Arctic regions, whilst its winter range touches pretty well every extensive coast-line south of the Tropic of Cancer; third, the Curlew Sandpiper (*Tringa subarquata*), which breeds on the highest North Polar land, and extends its winter flights as far as Australia; fourth, the Knot (*Tringa canutus*), which breeds within the narrow compass of the North Polar Basin, and in winter wanders to Australia, New Zealand, Cape Colony, and Brazil; fifth, Bonaparte's Sandpiper (*Tringa bonaparti*), which breeds in Arctic America, and in winter reaches Patagonia and the Falkland Islands; sixth, the Red-throated Stint (*Tringa ruficollis*), breeding in North-eastern Siberia, but how far north is not precisely known, and in winter extending its southern flight to Australia. Two of these species

(*Calidris arenaria* and *Tringa bonaparti*) are even reputed to have bred in the southern hemisphere, the former as far south as Lord Howe's Island, and the latter in the Falkland Islands, which, if true, strongly confirms a previous Antarctic habitat. At least twenty species of Charadriinæ birds from the Arctic regions visit Australia during winter; as many more probably reach South Africa and South America at that season, in addition to the sedentary species.

One may very naturally ask the reason for these vast extended flights.[1] Why are they undertaken for no object that man can determine? Migration is by no means a habit subject to caprice; nor do birds ever undertake it either in space or time without serious cause. I am of opinion that these long journeys, some of them reaching over 140 degrees of latitude, or nearly 10,000 miles direct, are the result of the transfer of these species from the North Polar Basin to the South Polar Basin,

---

[1] The suggestion made by Mr. Harvie-Brown (*Report, Migration of Birds*, iv. p. 71), that it is owing to the hardiness of constitution of birds bred in high Arctic latitudes that their fly-lines are so extended; and also to the annual overflow of enormous numbers pressing from behind and urging on to the south those in the van, needs little to refute it. The Knot (*Tringa canutus*) is given as an instance. Unfortunately for this ingenious theory, the Knot is bred in a high temperature, just in the very height of the short Arctic summer, and begins to draw south as soon as it can fly. It has, therefore, at no period of its existence any experience of the rigours of an Arctic climate; but, on the other hand, lives almost in perpetual summer or spring. Strength of constitution seems to have no effect whatever on the length of migration flight; the influences are of a far more deeply-rooted character.

during favourable intervals of climate. These birds extend their flight towards that ancient Antarctic habitat as far as they can find land free from snow on which to rest, impelled by hereditary impulse and inherited love of home. In those remote ages, when the Antarctic world was the favourite ancestral breeding-grounds, and the Northern Hemisphere their winter home, the flights would be just as extended as they are now. Birds are conservative creatures, and continue to follow old routes with much persistence, as we shall ultimately find. Of course we should only expect to find comparatively few birds doing this; and for many obvious reasons, most important of which is the small percentage of birds whose habitat is strictly Polar, and the vast remoteness of the time since Antarctic land was habitable; for the evidence is decidedly in favour of a very long-continued state of glaciation, owing probably to the absence of warm ocean currents and the presence of considerable areas of very elevated land. Consequently, the majority of species have been able slowly to adapt themselves to the conditions which render such extended flights unnecessary, or many others may have bred in lower latitudes of the Northern Hemisphere which did not suffer so much acute change of climate. For the sake of argument, these Polar birds of extended Migrations breed in the Northern Hemisphere, and winter in the Southern Hemisphere; although it must be understood that many individuals have ceased to perform the journey in its

fullest extent, but winter in various parts of the
Northern Hemisphere, where climatal conditions
are suitable.

Now if this were actually the state of things
during remote ages when the South Polar Basin
was a great breeding-ground for Waders and other
birds, it is only reasonable to expect that some
evidence is still left to us in support of our con-
jecture.   Such evidence fortunately is forthcoming.
There are many species of the CHARADRIIDÆ left
behind in the Southern Hemisphere, remnants of
that Great Exodus of byegone ages, some of them
very ancient relics indeed, as, for instance, the three
species of *Phegornis*, one of which (*P. leucopterus*)
is stranded on the Society Islands; another (*P.
cancellatus*) on the Paumota Archipelago, and the
third (*P. mitchelli*) on the Peruvian Andes.   These
birds are remarkable for their rounded wings, sedentary habits, and other peculiarities, and are acknowledged by one of the greatest living authorities on
this group, Mr. Seebohm himself, to be "the least
changed descendants of the ancestors of the Sandpipers."   Their presence on these Pacific Islands and
coast of South America appears to me to indicate an
ancient route of Migration across the Pacific Ocean
in the extreme east and west, from one Polar region
to another—a route which now is to a very great
extent discarded by existing species, although even
at the present day there is a considerable Migration
across that ocean by way of the Malay Archipelago
and Australia.   For, just as the line of present

Migration may in many cases indicate the direction of past Emigration, so may the present centres of Emigration denote ancient paths of Migration. More ancient evidence still of species stranded during Inter-Polar Migration may be found among the CHIONIDÆ and the THINOCORIDÆ, composed of aberrant species of CHARADRIIDÆ, isolated on various parts of the borders of the Antarctic Ocean.

Mr. Seebohm has very elaborately sought to show that the CHARADRIIDÆ is a North Polar group of birds, and seeks to explain the presence of some of these undoubted ancient forms as due to an exodus from the Arctic regions during glacial disturbance; but these isolated species seem to me to indicate South Polar origin; for it is only natural to expect to find the least changed ancestral forms nearest to the centre of dispersal, and this we certainly do not find in the northern hemisphere. Just as the Arctic ALCIDÆ have their more ancient prototypes, the Antarctic IMPENNES, indicating a remote Inter-Polar Emigration, so it appears to me the birds forming the groups CHIONIDÆ and THINOCORIDÆ indicate a South Polar dispersal of the CHARADRIIDÆ, and a less remote Inter-Polar Migration. There is other evidence to show that many Charadriinæ birds are decidedly more Antarctic than Arctic in their dispersal—not northern species at all, but eminently southern, such as the Stilts (*Himantopus*), the Oystercatchers (*Hæmatopus*), and the Lapwings (*Vanellus*).

The same ornithologist's Glacial Theory of Dis-

E

persal applied to the CHARADRIIDÆ,[1] is not supported
by geological evidence. First, no less than three
glaciations of the North Polar region are required
to account for the present dispersal and differentiation of this family of birds into genera and species.
But the evidence of more than one North Polar
glaciation during Tertiary time is absolutely wanting ; all that can be gleaned points to one Glaciation
only, which, as we have already seen, was Post
Pliocene. Secondly, the dominant line of Migration of these birds is unquestionably Inter-Polar,
which forces us to the inevitable conclusion that
the ancestral forms of this group of birds during
remote ages peopled a South Polar area. Another
fact strongly confirming Inter-Polar migration is
the comparatively cool temperature which appears
to be imperatively necessary for the majority of
these birds during the breeding season. It seems
hard to believe that such Polar birds could ever
become so thoroughly acclimatized to tropical zones
as to remain isolated in them for thousands of years,
as this Glacial theory of dispersal demands, when
an Antarctic Polar haunt was open to them, especially as we know that as soon as the North Polar
regions became once more habitable a great exodus
to them commenced, probably as South Polar
haunts became glaciated. Again, it appears to me
that too much reliance is placed on the various
reputed routes, along which species are said to have

[1] Obviously suggested by *Evolution without Natural
Selection*, chap. I. (published in 1885).

emigrated at the close of the Pliocene age. There
is much evidence to show that the following of
these routes was highly improbable, if not impossible, simply because the land area of the north
temperate zone has undergone vast change during
Tertiary time, and must have had a very different
aspect from what it now presents.

Let us now turn to more recent testimony. In
the Falkland Islands *Charadrius modestus* breeds,
and goes north to winter as far as Uruguay; whilst
an allied form, *C. modestus rubecola*, breeds on
Tierra del Fuego, and passes as far north as Chili
during winter. The fact that this species has become segregated into two races in so narrow an
area seems to suggest a long residence in this
region. In the same group of islands we also have,
in Tierra del Fuego, *Charadrius sociabilis*, with
the same distribution as *C. modestus*, and *Scolopax
frenata magellanica*, breeding in the Falklands, and
passing as far north as Paraguay to winter. In
Chili we find *Himantopus brasiliensis*, with a considerable northern migration during winter; also
Rhynchæa semicollaris, which in summer extends
its range south to Magellan, and in winter north
to Brazil. In fact, all the species of the genus
Rhynchæa appear to me to be decidedly Antarctic.
Their present distribution seems to admit of no
other conclusion. Their presence in Patagonia,
South Africa, Australia, and India (probably by
way of the Malay Archipelago, Siam, Burma, and
Lemuria), rather shows an Antarctic emigration than

a centre of dispersal in India, due to Post-Pliocene Arctic emigration, which spread by no conceivable way to South America! In the Snares and Chatham Isles *Scolopax auklandica* is a resident; in New Zealand we find *Ægialitis nova zelandiæ*, and *Charadrius obscurus*, the latter breeding on the mountains and wintering on the plains; whilst in Australia *Charadrius australis* may be given as an instance. Now, I think, after having seen that some species even at the present day migrate across the world from north to south, that others are isolated and sedentary in the Southern Hemisphere, having allowed their migratory impulses to lapse altogether, whilst others yet again breed as far to the south in the Southern Hemisphere as suitable land occurs, and pass north to winter, we are driven to the conclusion that Inter-Polar Migration amongst the CHARA-DRIIDÆ must have extensively prevailed in remote ages. It should also be remarked that none of these birds pass north of the Equator to winter—a fact which in itself seems to suggest a South Polar origin, whilst the number of Arctic species that winter south of the Equator is very considerable, and seems to show a lingering attachment to an ancient home.

Much evidence might also be given to show that the ANATIDÆ, or Ducks, and a portion at least of the Turdinæ (*Merula* and *Turdus*, or their common ancestors), have in past ages been Inter-Polar likewise. The HIRUNDINIDÆ, or Swallows, seem to be an Antarctic group which have comparatively recently become almost Inter-Polar in their migra-

tions, and would undoubtedly become entirely so
if milder climates again prevailed in North Polar
regions. We should then find many species of
this group inhabiting an Arctic area, with other
species left behind in southern latitudes, just as
we find to be the case with the CHARADRIIDÆ and
the ANATIDÆ at the present time. North Polar
conditions are now sufficiently favourable to allow
great numbers of Waders and Ducks to visit the
Arctic regions, although not sufficiently so for
many species of Swallows to visit them, and as a
natural consequence the latter birds are still the
most abundant in the Southern Hemisphere, whilst
the former birds predominate in the Northern
Hemisphere. One thing appears to be remarkable
about the migration of these Southern Hemisphere
species, and that is their comparative shortness.
This appears due to the very low southern breeding
area. Undoubtedly in the event of the South Polar
region once more becoming free from ice, the
migratory flights of these birds would be gradually
extended south, until they equalled in length those
of their Arctic allies. Inversely, we can understand
the migrations of Arctic species gradually becoming
shorter as glaciation sealed their Polar haunts;
culminating probably in a grand emigration of
many northern species to Antarctic regions, as we
have been endeavouring to demonstrate has actually
been the case in past ages.

In the preceding pages of this chapter I have
sought to impress upon the reader the fact that

Climatal Change, if it has been enormous and severe, has been gradual. We should naturally expect to find therefore the result of this reflected in present Migration. This is undoubtedly the case; for Migration is gradated in many various degrees both of space and time. It is by no means universally extended among species, or even, as we have already seen, among individuals of the same species (*conf.* p. 20). The flights were short at first, and were very gradually extended; varied likewise as to time and performance, as the seasons became shorter or longer. The evidence for this is abundant, convincing, and indisputable. The irregular appearance of many northern species in such southern latitudes as the British Islands, during exceptionally severe winters, admirably illustrates how under more pronounced conditions of persistent cold, Migration could be initiated. It is easy to understand how, if the winters were gradually to become uniformly and constantly more and more severe, these species, instead of paying only occasional and irregular visits south, would in the course of ages become regular migrants; and the habit or impulse to migrate, owing to the laws of heredity and a continual climatal cause, would become finally a deeply-rooted one. There can be no doubt whatever that this is very similar to what has actually taken place, not only throughout the varying climates of Præ-Pliocene ages, but during the lowering temperature of the Pliocene period itself, which, according to Professor Marcou and other scientists, was actually the beginning of the Glacial Epoch ; and yet again whilst that epoch was slowly pursuing its terribly majestic course along 200,000 years of time, with alternating mild periods at its inception and its close. We may instance among many others the Gray Phalarope (*Phalaropus fulicarius*), Ross's Gull (*Rhodostethia rossi*), the Glaucous Gull (*Glaucus glaucus*), the Iceland Gull (*Glaucus leucopterus*), the Little Auk (*Mergulus alle*), the Harlequin Duck (*Clangula histrionica*), Steller's Eider (*Somateria stelleri*), and the King Eider (*Somateria spectabilis*). Now all these species are either Polar or Arctic in their distribution, more or less stationary (in the sense of having but a slightly marked southern flight in winter), and only appear very irregularly in localities far south of their usual habitat, and then only when exceptionally severe winters in the north compel a southern movement. None of these birds remain long in the south ; they retire as soon as milder conditions prevail, and their movements although migratory are irregular and uncertain. They are examples of what may be termed Incipient Migration. Whenever climatal conditions become more severe, entailing absence of food, regular and more extended flights may become the rule among at least some of these species. We shall refer more particularly to this matter in a future chapter.

From species whose migratory movements are at the present time incipient, we will now pass to others in which they are decidedly regular, though small. Excellent examples of short migration flight

are presented in such species as the Rufous Warbler
(*Sylvia galactodes*), the Woodchat Shrike (*Lanius
rufus*), the Woodcock (*Scolopax rusticola*), and
the Golden-eye (*Clangula glaucion*).   Moderate
journeys are represented by the Great Reed Warbler
(*Acrocephalus turdoides*), the Turtle Dove (*Turtur
auritus*), the Crane (*Grus communis*), the Jack
Snipe (*Scolopax gallinula*), and the Mallard (*Anas
boschas*).   Amongst birds whose migrations are
decidedly long, we may note the Wheatear (*Saxicola œnanthe*), the Cuckoo (*Cuculus canorus*), the
Corn Crake (*Crex pratensis*), and the Greenshank
(*Totanus glottis*).   The most extended flights of
all known birds are performed by such birds as the
three species of British Swallows (*Hirundo rustica,
Chelidon urbica*, and *Cotyle riparia*), the Turnstone
(*Strepsilas interpres*), the Gray Plover (*Charadrius
helveticus*), the Sanderling (*Calidris arenaria*), the
Knot (*Tringa canutus*), and the Pectoral Sandpiper
(*Tringa pectoralis*).   The following table will serve
to demonstrate the gradated character of migration.
The mileage is approximate, and represents a course
almost due north and south ; but few, if any species,
however, travel so direct, so that the actual distance
traversed may be in excess of the figures on the
following page.

From the above facts I think that we may very
fairly make the following deductions.   First, that
incipient and short migration flight indicate movements to correspond with comparatively small
variations of climate necessary to requirements and

INCIPIENT	Waxwing Gray Phalarope Iceland Gull Ivory Gull Little Auk Steller's Eider	Mileage erratic; but normally from 1000 miles downwards
SHORT	Rufous Warbler Woodchat Shrike Pied Wagtail Stone Curlew Woodcock Black Tern Golden-eye	From 1000—2000 miles
MODERATE	Rock Thrush Great Reed Warbler Bee-Eater Turtle Dove Spoonbill Crane Lapwing Jack Snipe Mallard	From 3000—5000 miles
LONG	Wheatear Sedge Warbler Blue-headed Wagtail Nightjar Cuckoo Corn Crake Greenshank	From 6000—7000 miles
EXTENDED	Swallows Turnstone Gray Plover Whimbrel Knot Pectoral Sandpiper Sanderling Curlew Sandpiper Asiatic Golden Plover Hudsonian Godwit	From 7000—10,000 miles

consequent supply of food, together with gradual extension of range longitudinally, not essentially dependent on climate, and that these movements are the most recent in geological time. Second, that moderate and long migration flight are the result of a gradual extension of summer range, chiefly latitudinally, owing to modification of climate, with consequent increase of individuals, leading to an expanse of winter area, and are of much older initiation. Third, that extended migration flight implies a complete change of breeding-ground, more or less Inter-Polar, and is decidedly the most ancient of all.

In the present chapter we have glanced at the probable origin and descent of Birds; at the changes of climate they have experienced from the date of their evolution from a lower form; endeavoured to show how these changes have been brought about, and what the result has been. We have glanced at some of the most powerful causes of Migration, and traced the wonderful gradations of Migration Flight to their origin both in their mildest and in their acutest phases. We will now conclude by following in detail the migration of some single species, say from its Post-Pliocene glacial initiation to the present day, in order clearly to demonstrate Why the habit has been acquired, and How it is practised.

We will select the Spotted Flycatcher (*Muscicapa grisola*) for the purpose. It is one of our best known summer migrants, and one whose present geographical distribution admirably illustrates the phenomenon of Migration. When the Sub-Polar regions of the Northern Hemisphere last enjoyed a warm, almost semi-tropical climate—one of the mild periods of the Glacial Epoch—the Spotted Flycatcher inhabited in one unbroken area the Arctic woodlands from the Atlantic to the Pacific. Probably it was a resident species becoming partially nocturnal during the Polar night; food was abundant; its conditions of life were easy, and it multiplied apace, and became a dominant, firmly established species during the thousands of years that it dwelt in this Sub-Polar habitat. So matters continued until the slow precession of the equinoxes, in conjunction with increasing eccentricity of the earth's orbit, began to have a marked influence on the climate, and gradually the fair forests and the verdant plains were devastated by the ever-increasing cold. Age after age the Spotted Flycatcher was driven slowly south; summer after summer grew colder and shorter, the periods of Polar darkness more severe. At last matters became so serious that the birds began to leave their northern haunts in autumn, probably because their food became scarce as the various insects either retreated south or began to hibernate. Further and further southward these annual journeys had to be taken, until the Flycatcher at last found its way during winter into Africa, Persia, Arabia, India, China, and even the Philippines and the Moluccas. Summer after summer the belt of breeding-ground became wider and wider, and vast numbers of individuals became

separated from the rest of the species by the lofty
mountain ranges, the deserts, and other physical
barriers, which would effectually arrest a forest or
woodland haunting species.   More and more severe
became the winters, longer and longer ; the glaciers
descended   lower   and   lower, exterminating   or
driving before them all living things.   At last the
Spotted Flycatcher, or the form which then represented this species, came to be divided into two
enormous colonies—an African one and a Chinese
one—the individuals of each being completely
isolated from each other, summer and winter alike.
During the ages that this state of things continued,
the Flycatchers became segregated into two species,
owing primarily to the absence of any intermarriage ;
the eastern race became smaller, the tail shorter,
and the breast-streaks broader ; or the western race
became larger, with a longer tail and narrow breast-
streaks.   It is almost impossible to say which form
now most closely resembles the ancestral species, but
such are the present differences between the two races
known to ornithologists respectively as *Muscicapa
grisola* (the Western and British form) and *Muscicapa griseisticta* (the Eastern form).   Such was the
state of things at the close of this Inter-Glacial Period.

Then came the gradual immigration north again
as precession and lower eccentricity initiated a
milder climate.   Age after age the journey in the
spring became longer.   Certain routes to and fro
became to be recognized highways of passage ; and
so imperceptibly did the northern breeding-grounds
expand, that the birds became regular migrants,

looking upon the movement north to higher and cooler latitudes each spring as an undertaking never to be missed.   Warmer and warmer became the southern haunts, stimulating and widening migration flight to the cooler temperatures prevailing near the edges of the retreating glaciers, where a suitable breeding climate could only be found.

Let us confine our attention solely to the birds that bred in the British Islands.   In the Præ-Glacial ages this area formed part of Continental Europe; a rich and fertile corner, abounding in insect life, full of haunts the Flycatcher loved. After the banishment of its race and the exile of its ancestors in Africa, the northern journey at first did not extend further than the edges of the glaciers on the Mediterranean coasts of Europe.   But as these disappeared, and a warmer climate began to prevail in higher latitudes, the annual summer flight was increased.   Every century the northern breeding range had increased; creeping slowly across France; higher and higher with the growing vegetation; nearer and nearer to the haunts of old. During the slow gradual elevation and submergence that isolated Albion from the rest of Europe during Post-Glacial time, the regular spring journey across the sea became wider and wider; but with the intense and inherited love of home in their tiny breasts, the individuals that were born and bred in this district never failed to return each year. For 60,000 years or more has this species now crossed the sea, returning every season, not only to our islands, but each pair of individuals as long as

they live come back to the exact locality of their
previous nests. This long journey, gradually growing longer and longer during thousands of years, until
it is now at least a thousand miles in length, has
grown to be a deeply-rooted custom sanctioned by
the practice of ages of experience and need, and
looked upon now as part of the Flycatcher's very
existence !

The above interesting and thoroughly demonstrable instance shows two very important ways in
which Migration has been initiated. First, by a
gradual movement south during a short winter,
growing more extended in space as the climate
became more severe, and more prolonged in time
as the winters became longer. Second, a gradual
movement north again with the return of milder
climatal conditions. In both movements, however,
the first journeys undertaken were probably of very
small extent, and slowly became longer as the
summer area of distribution became wider. During
this period of Glaciation, bird-life was utterly
banished from all northern and many temperate
lands. During the past 60,000 years a general
avian exodus has been in progress. All the hardier
birds have emigrated back as far north as they can
live, and many have become sedentary in the more
temperate areas, but the winters still continue too
long and too cold for the majority of insectivorous
birds breeding in those areas, with the inevitable
consequence that Migration among them is the
almost universal rule.

CHAPTER III.

THE PHILOSOPHY OF MIGRATION.

Wings of Migratory Birds—Plumage of Migratory Birds—
Moulting—Order of Migration in Autumn—*Avant-Courières*
—Migration of Young Birds before Parents—Order of
Migration of Adults—Order of Migration in Spring—Time
of Migration Flight—Punctuality of Migrants—Amount of
Sociability and Gregariousness in Migrants—Weather-bound
Migrants—Duration and Progress of Migration—Speed of
Migration Flight—Rapid Flight of Dotterel in Spring—
Altitude of Migration Flight—Influence of Bright Lights
on Migrating Birds—Migrants at Lighthouses and Light-
ships—Effects of Reflected Light—Advantages derived
from Lofty Flight—Possible Use of the Balloon in studying
Migration—Cries of Birds on Passage—Effects of Wind on
Migration—Influence of Temperature on Migrating Birds.

THE two preceding chapters have been primarily
devoted to the biography, antiquity, and origin of
migration ; it now becomes necessary to deal with
the actual phenomenon itself. We have seen when
and how migration had its origin, now let us trace
the various conditions under which it is practised.
Flights of such enormous magnitude are not lightly
undertaken, and many things have to be considered
during their progress. Even before they are commenced certain preparations have to be made, and
many, and varied, and complex are the influences
inseparably associated with these wonderful journeys.

In the first place birds must be in good condition before their long pilgrimage can be undertaken with any degree of comfort or hope of success. A flight of many thousands of miles requires much physical exertion; the fatigues of such a long journey can only be undergone with success, and the perils reduced to a minimum, by birds capable of performing the distance in a reasonable time, and withstanding the enormous constitutional strain. It is a rule almost without exception that the wings of migratory birds are long and pointed —a form best adapted not only for sustained but rapid flight. It may also be remarked that the birds which cover the longest distances have the most pointed wings. The Turnstone, with its enormous migration flight of ten thousand miles or more, has a long pointed wing, the first primary being the longest, and this is the rule throughout the CHARADRIIDÆ, the only exceptions being of species whose flights are very short or dispensed with altogether, as in *Phegornis:* the Swallow's wings are equally pointed. The laws of use and disuse determine the shape of the wing of a bird. The wings of a bird vary according to the amount of flight indulged in by their owner. Birds that fly much have pointed wings; birds that fly little have short and rounded wings; whilst birds that rarely fly at all eventually degrade into species utterly incapable of flight, as was the case with the now extinct Great Auk, and is the case with the still existing Antarctic Penguins. Long pointed wings,

however, do not necessarily indicate migratory
habits, for many sedentary birds possess them; but
these species, as for instance many of the Raptores,
some of the Terns, many Humming Birds, and
many Swallows, all depend largely on their wings
for obtaining food. Let us glance at a few instances.
The wings of a Goldcrest or a Willow Wren are
long, flat, and pointed, best adapted in every way
for migration flight; but the wings of a Wren or
of a Dipper are short, more concave, and rounded,
a shape most ill-adapted for long-sustained flight.
All, or nearly all, the TIMELIIDÆ, most of the
PITTIDÆ, many of the PARADISEIDÆ, and hosts
of other sedentary tropical Passeriformes are
remarkable for their feeble, concave, short and
rounded wings.

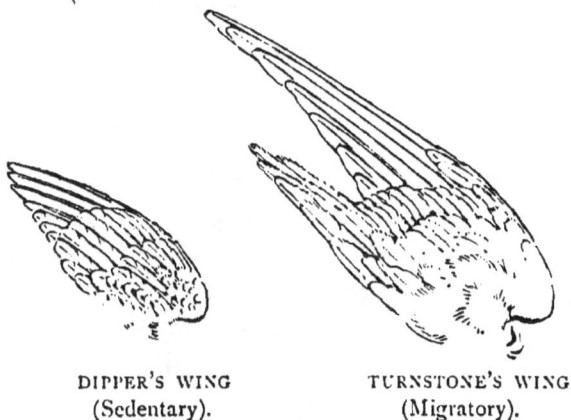

DIPPER'S WING TURNSTONE'S WING
(Sedentary). (Migratory).

This variability not only extends to the feathers
of the wing, but includes corresponding development of the bones which support it and the muscles
with which it is worked. That the shape of the

F

wing is correlated with the habits of the bird is very
significantly proved by the fact that in Arctic and
temperate regions where migration is most practised,
there are few birds with the short rounded concave
wings of the sedentary birds of the Tropics, where
migration is little practised. It should also be
stated that the plumage of migratory birds is
generally close and compact; a loose and fluffy
dermal covering is seldom worn by any but the
most sedentary birds. We may thus fairly begin
with the fact that birds of passage are both in the
formation of the wing and in the texture of their
plumage the best of all birds adapted for flight.

We will now advance a step further in our
subject. The feathers of a bird are renewed—in
other words moulted—as every naturalist knows,
at stated periods. These periods vary a good deal
according to circumstances, the only one of which
we need stay to discuss being the period which
seems correlated with migration. In many migratory birds, certainly in all those whose flights are
the most extended (as in the CHARADRIIDÆ), a
complete double moult takes place, in spring and
in autumn. In all others a change is undergone in
autumn, with the exception of the birds comprising
the LANIIDÆ (Shrikes) and the HIRUNDINIDÆ
(Swallows), in which the plumage is renewed in
early spring. These latter exceptions may be easily
explained. The birds in both these families appear
to belong to the Southern Hemisphere, and have
only within comparatively recent times extended

the area of their distribution northwards, so that
their moult which is performed (by the Swallows)
in February and March is really normal, that season
being the autumn of the Antipodes. The more
northerly dispersed Shrikes moult later, in our
spring in fact; but this is only the retention of an
autumn moult performed during bygone ages in
their ancient southern home. The great majority
of migratory birds therefore moult before they
migrate, either in spring, or in spring and autumn;
but some few, as, for instance, the Common Sandpiper (*Totanus hypoleucus*), moult in autumn in
their winter quarters just after their migration is
over, although in spring the change occurs before
passage, as some sort of compensation for the
irregular circumstance. No bird of single moult
ever migrates in autumn before changing its flight-
feathers, and the change is effected just previous to
the journey. Normally, and almost without exception, the moult of the quills, whether single or
double, comes to an end just before the time of
actual migration arrives; although some birds, as,
for instance, several species of Geese (*Anser*), begin
to walk from inland breeding-places *en route* to the
coast or great river valleys, changing their dress
as they go, whilst others frequently moult many
small body-feathers as they migrate. It must,
however, be borne in mind that the change of
quills is completed before the actual flight portion
of the bird's migration is commenced. During the
period of moulting, birds keep very close to their

haunts; they are songless, almost silent, skulk low
in the cover, or, as is the case with aquatic birds,
repair to the sea and open waters where the greatest
amount of safety is to be found. They are well
aware of their helplessness, and consequently keep
very close and quiet. But by the time the new
dress and the bright clean shaft-feathers have been
acquired, which give them a new lease of aerial
existence, migratory birds change their habits
considerably. The hereditary impulse to migrate
grows stronger and stronger within them, an
impulse whose principles must be obeyed as
inevitably as the impulse of sexual passion, and,
well equipped with a new covering of feathers and
a new set of flight-quills, the journey is commenced.

It must not be supposed, however, that all
individuals start off together; the order of precedence appears to be observed with as much
jealous strictness as at the courts of human kings
and princes. The Order in which the individuals
of a species migrate is influenced to a very great
extent by the progress of the moult, especially in
autumn. The quickening impulse to migrate in
autumn is for a time at least subservient to the
parental love of offspring; although as it reaches
its highest stage of development it appears even to
become stronger than parental love. Witness the
Swallows and Swifts deserting occasionally late
broods in autumn; leaving them in their nests
helpless and to starve. Only last autumn (1891)
I knew of a brood of three young Barn Swallows

left behind by their parents to perish, after lingering
by them until the early days of November, when
they were almost able to take care of themselves!
But with birds that from some cause or another
have not been able to breed, or have accidentally
lost their eggs or young broods, no parental instinct
exerts any restraining influence, and the desire to
migrate often becomes so prematurely strong that
they begin to leave their summer quarters in some
cases even before their moult is absolutely completed. These birds are the pioneers—the *avant-
courières* of the migrating army; the guides of the
inexperienced, the heralds of the advancing host.
Strange, impossible as it may seem, it is nevertheless true that the Young Birds—the birds that
have never travelled before—are the next in order
to migrate. In their case the impulse to migrate
must be entirely hereditary, or nearly so; what little
external influence incites them probably exists in
the fact of seeing the *avant-courières* depart.
These young birds are in the normal course of
things the first to be in a position to migrate; they
travel in their first plumage, and consequently are
ready to go as soon as they can fly. That they do
not tarry long at their birthplace after this time
arrives is proved by the fact of their being seen in
latitudes far to the south of where they were born a
week or so after we know that their flight-feathers
reached maturity. Again, young Knots (*Tringa
canutus*) and young Gray Plovers (*Charadrius
helveticus*), bred in some of the highest Polar dis-

tricts, are occasionally met with on the British coasts with particles of their downy nestling plumage still adhering to some of the feathers, after a flight of at least 2000 miles. It must not be supposed, however, that because the impulse to migrate is inherited from their parents, the *ability* to do so is equally hereditary. That has to be acquired; the road has to be pointed out by the more experienced guiding birds, and the long, often circuitous, route has to be *learnt* by the experience of not one but many annual journeys to and fro. Wonderful therefore as this order of migration is—and that it is a fact is proved by overwhelming testimony at every station where the passage of birds has been studied—there is nothing abnormal about the proceeding; and we see that the odd restless old birds that migrate before the rest, incite the young birds to start, and render the important service of showing those young and inexperienced the way. A week or so after the young birds have left, the adult males begin their migration, having got over the moult a little earlier than the females, the latter being delayed somewhat by maternal duties, so that their departure is a little later still. The rear of the great migrating army is brought up by the birds that from various causes have been either prevented from starting with the rest or delayed on the way, by such accidents as damaged flight-feathers or maimed and wounded limbs. Astonishing as this fact may be, it has been remarked and verified

repeatedly by Gätke whilst studying migration for
half a century on the tiny island of Heligoland,
one of the best adapted spots in all the world for
the purpose. The order of return in the following
spring is partially reversed. Then the adult males
take precedence; the females travel next in order;
the birds of the year follow these, and last, as usual,
come the weakly and the wounded, following on
in the wake of their more vigorous and healthy
comrades, like so many camp-followers and disabled
travellers in the rear of a marching army!

The daily time of migration flight varies a good
deal. Some species more habitually migrate by
day, and may frequently be watched on their journey
north or south, as the case may be, all the time that
the sun is above the horizon. The great majority
of birds, however, migrate by night, or if they do
pass by day, it is above the range of human vision.
Many species, however, will continue their flight
along certain routes after the sun has risen; yet, on
the other hand, numbers prefer to rest for the day,
provided they are on land, wherever they may
chance to be, passing on again with the recurrence
of darkness. Every observer of birds must have
remarked these interesting facts over and over
again. At sunset one day not a bird of a certain
species can be found anywhere; at dawn the next
the place may swarm with them. On the other
hand, he may see a species abundantly one evening,
and search for it in vain by daylight on the following morn. During spring and autumn, odd birds,

or little parties of birds, may be seen from time
to time during the day flitting along; or suddenly
a district may swarm with a species or even a
number of species, which remain loafing about for
the day, but not one will be left by the morrow.

The punctuality of arrival of birds either at their
summer or winter quarters, or at various points *en
route*, is nothing less than astonishing. Taking
into account the length of the journey they have
come, and the consequent number and variety of
possible causes of delay, the best kept time of the
crack expresses, or the passages of the fastest
steamers that plough the ocean with a space-annihi-
lating speed, absolutely suffers by comparison!
A one hour late in the 200 miles' run of an express,
or twenty-four in the 5000 miles' voyage of a
steamer, is certainly, all things considered, a far
worse record than the one day late in the 5000
miles, or the couple of days in 10,000 miles' flight
of a bird, at the mercy of countless contingencies
neither the train nor the boat have to battle with.
And yet this is the simple statement of facts.
Migratory birds may be looked for almost to the
day, as any one can prove by keeping a record
during a series of years. The arrival of sea birds
at many of their breeding-places is so regular, that
it forms a date in the calendar of men most concerned in the event. Of course, the date of arrival
varies a good deal with different species. Some
species migrate earlier, some later than others; the
migration flight of each being regulated by various

influences more or less important, the chief of which is undoubtedly the supply of food. It is also worthy of remark that the birds migrating earliest generally stay the longest; whilst those that are the latest to arrive are the earliest to depart.

The amount of sociability in birds whilst on migration is not only very interesting, but very variable. Some species are solitary pilgrims indeed, coming and going in a very exclusive manner; if crowded together from some unusual cause *en route*, taking the first opportunity of separating again, even from the companionship of their own kindred. The Woodcock (*Scolopax rusticola*) may be instanced as a recluse; the Nightjar (*Caprimulgus europæus*) and the Cuckoo (*Cuculus canorus*) as others. Who has ever witnessed any gregarious or even social tendency in the migrations of these birds, always excepting the occasional crowds or "rushes" that are due entirely to accident, as we shall shortly learn? Other birds, however, are remarkably gregarious during migration, many species only becoming so at such a time, and even in species that are always gregarious the habit is more intensified during Flight. Other birds are not only gregarious but social also, and it is not unusual to find migrating flocks composed of parties of several species: witness the great congregations of the Swallow tribe, and of various Waders and Ducks at these periods. Many odd birds will also, from time to time, join a flock of some other and often very distantly related species, apparently for the sake of company on the road. Other birds, especially those that pair for life, invariably migrate each in the company of its mate; and in such species as pair at their winter quarters, the vernal migration is always performed together, whether the individuals gather into flocks or not.

It very often happens during the course of the migration period, which, roughly speaking, lasts about three or four months in spring, and again in autumn, that a spell of bad weather, or the long persistence of unfavourable winds, will arrest the flight of migrants. Throughout the area of meteorological disturbance, Migration practically ceases for the time being. During the prevalence of unfavourable weather, however, great numbers of birds, sometimes of many different species, gradually accumulate, like so many wind- and tempest-bound vessels in a quiet bay, all delayed and waiting the first favourable moment to push on again. As soon as the weather changes the hosts of birds pass on; and it is to this cause we may chiefly attribute the spasmodic rushes of a species, or many species together, which so often occur, especially during the stormier and most unsettled months of the migration season. Owing to this, many eminently solitary birds are sometimes observed to arrive at a given point in company, but they invariably separate as the great influx spreads. These rushes are most frequent in autumn, not only because storms are more prevalent then, but because the number of migrants is greater than in spring.

The Migration of most, if not all, species may
be roughly calculated at from one to two months
each way, and has its regular phases of intensity.
In some nomadic species it lasts much longer,
as we shall learn in a future chapter. First, the
migration of a species is marked by the advent or
departure of a few stragglers, the most venturesome
and restless individuals; then the flight becomes
more vast until the culminating point is reached,
perhaps in two marked rushes, after which it ebbs
pretty much as gradually as it flowed. The phases
of intensity are more marked and sharply defined
in some species than in others, and are due probably
to various local causes. It must not be supposed,
however, that the period during which the migration of a species lasts, is actually taken up in direct
flight. The time is taken up in draining the
summer or winter area, and is longer or shorter
according to the width of latitude cleared. Where
the breeding area of a species extends over 3000
miles of latitude, with its consequent great variation of climate and date of season, the migration
will be prolonged; birds starting from the most
northern localities months before they do from the
more southern ones. Where the breeding area of
a species is narrow the migration is short, because
most of the birds start at the same time. It is the
same with the winter quarters. Where they extend
far to the south they are filled more slowly than
where they are restricted to narrower zones. The
evidence that birds leave their summer or winter

quarters, and never stay until they reach their
destinations, amounts practically to nothing, for
the instances of continued flight are exceedingly
rare. Birds travel by stages, staying here or there
on the way to feed or rest, so that it is an extremely
difficult matter to arrive at the rate of speed of the
actual migration flight. That some birds can fly
amazingly fast is unquestionable; that they habitually do so on migration is not supported by facts.
The only stage of their journey which it is of vital
importance for them to get over quickly, is that
across the sea ; some birds, however, have no sea
to cross at all. Over land migrating birds appear to
fly at a moderate speed and with great persistency,
like a trained pedestrian who has set himself a long
task of endurance, plodding along in a steady but
continuous manner. We can only make the wildest
guesses at the time occupied by individual birds in
reaching their summer or winter quarters; consequently it is equally impossible to give a time-log
of their route. Probably migrating birds do not
average more than 300 miles per day, during
their journey north or south; but certainly birds
travel quicker north in spring than they do south
in autumn. Of course, this does not adequately
represent the velocity of their flight between the
stages. Some birds can fly amazingly quick;
Swifts probably can and do attain a speed of nearly
200 miles per hour. There is, however, one
instance by which we can form some idea both
of the rapidity of flight and the briefness of

migration time, if necessity demands a high rate of
speed and a short time for transit.   The Dotterel
(*Eudromias morinellus*) breeds on the tundras of
Arctic Euro-Asia, and winters in Africa, north of
the Equator.   Its spring migration is late and
rapid, and as the bird is scarcely ever seen in intermediate localities during this season (Heligoland
records but few in May), we are forced to the conclusion that this enormous flight of quite 2000
miles is performed without a rest, and between
sunset and sunrise.   If the Dotterel were to start in
the evening gloom from its African haunts, say at
seven, it would reach the moors of the Arctic
regions, by flying 200 miles per hour, about five
the following morning—a record of speed that
makes the highest pace of our " Flying Scotch-
man," " Wild Irishman," or " Dutchman " appear
but the creep of a snail by comparison, and of
astounding endurance, which may well fill us with
genuine admiration and wonder.   As is usual, the
flight is slower in autumn, and then the Dotterel is
observed on passage in the ordinary way, crossing
Heligoland in August, passing through Germany
in September, and Malta in October and November,
but always very rare over the British Islands at that
season.

Equally interesting, and perhaps even more important than the Velocity of flight, is the Altitude
at which it is performed.   In my opinion, the vast
importance of altitude in migration has never been
recognized by naturalists.   I will even go so far as

to say, that without a considerable altitude the
migrations of many birds would be simply impossible; unless we were to attribute to these
creatures mysterious and supernatural powers of
perception, which would not only be most unscientific, but excessively absurd.   That most birds
fly unusually high during migration is, I think, an
unquestionable fact.   Witness the vast height to
which Swallows and Swifts will soar just previous
to departure, or the startling suddenness with which
migrating birds will drop perpendicularly from the
sky when their flight has been unexpectedly arrested
by meteorological influences.   Other proofs are to
be found in the fact that migrants rarely strike
against lighthouses except during spells of sudden
darkness, due to fogs or clouds, which compel them
to seek a lower altitude.   Again, birds may be
actually observed migrating at vast heights.   Gätke
records Rooks on passage flying so high that they
looked like dust, and were only recognized by their
cries.   Of course, the size of large birds assists us
to distinguish them at an altitude of many thousands
of feet, especially those that habitually migrate by
day; but small birds may wing their way entirely
undetected at such enormous heights.   Again,
migration, as we have already stated, is mostly
undertaken at night when birds cannot be seen at
all, although their cries may be repeatedly heard
obviously at a vast height as they wing their way
across the starry skies.   As another instance of vast
altitude during migration, I may mention the fact

of Cranes being observed on passage at a great elevation crossing over the Pamir Plateau in Central Asia, a district which is upwards of 16,000 feet above the level of the sea. Much evidence might also be given of migration extending through lofty mountain passes, which it is only fair to presume are reached by horizontal flight, rather than by vertical flight from the plains below. Some of the most elevated land (not actually mountain summits) in the whole world is known to be the pathway of migratory birds.

By far the most important benefit derived from following an extreme lofty course is undoubtedly that of increased range of vision. The higher a bird flies the further it can see, the more extended becomes the visible segment of the earth's sphere below it. It is probably almost entirely due to its aerial existence that a bird's powers of perception and knowledge of locality are so acutely developed. There can be little doubt that the habit of constantly viewing the country at various heights, in all kinds of weather, and under ever-changing atmospheric influence, endows a bird with a knowledge of topography that seems to a terrestrial animal like man as little short of marvellous. The lay of the country can be seen at a glance from an aerial point of observation; and it is this well-known fact that leads a hunter lost in the forest to climb the nearest tree to ascertain his bearings. Each migratory bird must have a wonderful knowledge of the topography of its own particular routes, aided

by its marvellous power of memory and keenness
of sight. I would suggest, however, that the migration flight reaches its highest altitude when passing
over seas. These offer no landmarks, no bearings,
nothing that may serve as a guide; consequently,
the line of flight rises to a sufficient altitude to
enable the bird to bridge the passage with its keen
powers of vision. In confirmation of this, I may
remark that in no part of the world do any regular
migration routes cross seas too wide to be bridged
by the eye of a bird flying at a sufficient altitude.
The amount of the earth's surface within the
compass of a bird's vision flying at any known
altitude is easily calculated by spherical geometry;
and, putting the average altitude at the very low
estimate of from 1000 to 5000 feet, this will give
a wide enough area for all practical purposes. That
it is vastly exceeded, probably four- or five-fold, is
unquestionable. From this lofty course the earth
and sea for miles and miles will spread out below
these migrating birds in one unbroken panorama;
all the old familiar landmarks will be readily seen,
each peculiarity of coast, already known so well by
experience, descried. All the dangerous or un-
suitable localities, all the favourite places of call,
which years of former experience have taught them
to avoid or visit, will be displayed below them in
one broad and ever-changing expanse. This tendency to a lofty flight amounts almost to a passion
as the time of migration approaches, and terrestrial
birds that have kept close to the bushes or the

ground all the summer, suddenly become aerial, and mount upwards to pursue their way to distant lands. The majority of species appear to fly low as they near land; but others, as for instance the Wood-cock, keep at a considerable altitude until the land is below them, when they drop suddenly down into the nearest cover.

There is another matter which bears very importantly on the altitude of Migration Flight, and that is the singular influence of bright lights on birds during passage. As every observer of migration knows, the various lighthouses and light-vessels that stud the seas and coasts across or along the routes of migration are frequently the centre of attraction of various pilgrim birds, most especially during a spell of fog or haze, or a few hours' prevalence of cloud. The question arises, Why do birds thus make for these fiery points of attraction, a proceeding which only too often ends with fatal results? It is out of no mere curiosity or desire to examine an unfamiliar object, for the visits are conditional and exceptional, whilst to many of these birds lights must have been common enough throughout the preceding summer. The most probable explanation is, that the sudden appearance of fog or the drifting of cloud-banks between them and the earth has caused them to lose their bearings, to flutter aimlessly down into a lower stratum of air, and then to make for the nearest object at all likely to guide them to a place of safety. The brilliant lamps of the lighthouse are too often the

only visible place for which to steer. An inherent
knowledge teaches them that light leads to safety.
Indeed, it is by no means improbable that reflected
light acts as a great and important guide to migratory
birds. Now, the most important routes of migration
are near or over the great water-ways of the world
—down river valleys and coast lines; or along the
direction of mountain ranges, whose summits are
usually more or less covered with snow in autumn
and spring. Water and snow reflect light to a
very great extent, and it is easy to conceive how
a bird could gain some considerable knowledge of
its general course by following the gleaming expanse
below. The tide of migration flows high on moon-
light nights; starlight is also favourable to its
progress. Migration is discontinued during an
overcast sky, when the moon and stars are shut
out from the earth by a fog or cloud-veil, as is
abundantly proved by the promptness of migrants
to visit the lighthouses when the earth is for the
time hidden in gloom, and the readiness with which
the pilgrimage is resumed as soon as the heavens
are clear again. If the fog, however, is of a local
character only, migration is often continued above
it, because the earth area which it conceals is not
sufficiently large to cause birds to lose their bearings. Thus the seas, and lakes, and rivers, and
snow-capped mountains, glistening in the cold white
light of a brilliant moon, or gleaming gray yet full
of reflected lustre from a star-lit sky, serve as so
many flashing scintillating guides or steady glow-

ing beacons to the birds flying on, and on, and on
above them !

We may allude to several other possible advantages that may be gained by a lofty flight. The
mere mechanical labour of flight is rendered much
easier of performance in the more rarefied atmosphere of these lofty regions of space. This is a
matter of great importance to species whose fly-lines
are very extensive, for they are thus able to fly
longer with less fatigue, and *quicker* than in the
lower and denser atmosphere—two things very
essential to the successful performance of their
journey. The currents of air at a great altitude
may also be more uniform and favourable, whilst
lower down, the air stream may be blowing in a
direction quite unsuited to Flight. Another advantage of a lofty course is the greater immunity
from enemies, a very important item in migration,
as we have yet to learn. Far up in the sky, especially when shrouded in gloom, the way is free from
danger; and small birds, even in daylight, are not
readily discerned at any great height above the
earth. One other advantage to diurnal migrants
especially is the lengthened period of daylight.
The valleys are hidden in gloom long after the
mountain tops continue to reflect the light of the
setting sun; and these lofty summits catch the dawn
long before it reaches the lower country. I have
already alluded to the possible use of the balloon
in studying Migration (see *Idle Hours with Nature*,
p. 217), and here I would seek to emphasize my

previous suggestion. That a captive balloon floated above some spot where migration is notoriously prevalent, as for instance at Spurn Point on the Yorkshire coast, in the Wash, on the Sussex Downs, or better still, over Heligoland, would result in priceless information concerning the annual movements of birds, is absolutely certain. The more I study Migration, the more I feel convinced that it is a nocturnal drama of the air, and that only a faint conception of its wonders can be formed from terrestrial scrutiny. I shall be happy to assist in any such investigation. Will the British Association for the Advancement of Science, which has already done so much for the elucidation of the Pilgrimage of the Birds, still further lend its all-powerful assistance towards establishing such aerial posts of observation?

The various notes uttered by birds on migration must not be forgotten. Some species are much more noisy on passage than others, and many birds keep up quite a chorus of cries as they wing their way along. That these notes serve the important purpose of keeping migrating flocks together seems unquestionable. The cries of birds may be heard repeatedly at night during the two seasons of passage, as flock after flock of migrants crosses over the darkened sky. These notes may also serve as guides to the young and inexperienced, keeping them to the true course; whilst the birds that fly in silence may be kept in touch of the route by hearing the cries of more noisy species.

Geese, for instance, are noisy during Flight; so too are Waders. Crows, however, especially when migrating by day, are silent birds; but Larks, whether on passage by day or by night, are incessantly calling to each other, and very frequently burst out into song the moment they reach land during daylight. Many a time the only sign on earth that a great migration of birds is in progress, is by hearing the varied and oft-repeated notes, sounding faintly from on high as the armies of birds pass on in the darkness overhead.

A few words on Wind and Temperature will bring these meagre remarks on the philosophy of a great subject to a close. There can be no doubt that wind has a great influence on the migratory movements of birds. It is not every wind that is favourable to Flight; tempests arrest migration almost entirely; adverse winds retard it; but a gale, if from a suitable quarter, often seems more to assist it than otherwise. Strong head winds are always avoided if possible, for the obvious reason that a long-continued flight would be made with a maximum of labour and a minimum of progress. A wind blowing directly behind is also unfavourable to Migration Flight, for the cold current blows up through the plumage and chills the body of the flying bird. The favourite wind is a shoulder wind, what in nautical parlance is known as a "beam wind," or a wind blowing more or less obliquely across the line of flight. The direction of flight is usually within three or four points of the wind.

A southern migration is therefore best performed, not as some readers might imagine with a north wind blowing directly behind the travellers, but with a south-easterly or north-westerly wind blowing obliquely across their path. A northern migration is best advanced by north-easterly winds. Very light head winds are often favourable to migration; and in rarer instances migrants pass to windward with a gentle and warm breeze. Birds are very careful in their choice of wind, and nothing retards migration so much as contrary currents, the little travellers often waiting for days until a favourable breeze springs up, and they can renew their flight. Sudden gales from adverse quarters will frequently blow migrants hundreds of miles out of their proper course with fatal results, and occasionally bring them to countries which they would never normally have visited.

The influence of Temperature on migrating birds has not yet been studied sufficiently well to furnish much reliable data on which any very important generalizations or conclusions may be based. That it has important effects on Migration Flight can scarcely be doubted, and it is sincerely to be hoped that observers when recording migratory movements will pay due care to this portion of the subject. Rises and falls of temperature are evidently very important impelling causes of migration, especially of nomadic migration. It is even possible, if we allow ourselves to roam for a moment into the speculative realms of theory, that the gradual rise of temperature experienced as the several zones of the earth are crossed from north to south, may serve as some guiding influence, attraction, or gentle impetus to birds in search of warmer lands in autumn ; just as the fall in temperature experienced in a passage from the Tropics to the Temperate or Arctic zones might act inversely in spring. Birds migrate in spring to regions where the temperature is suitable to the vital function of Reproduction—and it is an axiom in ornithology that birds in the warmest areas of their distribution or habitat select the coolest areas of temperature in which to breed. In autumn, birds are in quest of higher temperature, or rather retreating from regions where the temperature is rapidly falling. That birds, like many other animals, may be able to foretell the near approach of barometrical disturbance, that this finely-adjusted faculty is of vital importance to such mighty travellers, seems so palpable that we can scarcely question its truth, although in the present state of our knowledge we fail to see how or why. Mr. J. A. Allen has remarked that in America the migrations of various birds precede storms or sudden falls of temperature often only by a few hours, thus avoiding these disturbances by keeping ahead of them. Great migratory movements have repeatedly been observed coincidently with favourable barometric conditions, and to cease with the reverse. We shall have occasion to allude to the influences of Temperature on Migration Flight again.

CHAPTER IV.

ROUTES OF MIGRATION.

THE next portion of our subject with which it
becomes necessary to deal, embraces the various
Routes that are followed by migratory birds. These
highways of migration are both numerous and well
recognized; many of them are extremely complicated. Probably the fly-line of no two species is
exactly the same; and the variety of route seems
almost as excessive among individuals as among
species. Each appears to follow a route which

past experience has taught it to be the best for its
own requirements, or the countless contingencies
and necessities of migration have compelled it to
learn. Consequently, we come across many apparent
anomalies in studying the fly-lines of birds ; we
shall find them intricately interlaced and even
crossing each other at right angles ; running
parallel but reversed by certain species according to
season ; turning without any apparent cause. If
it were only possible to focus the Palæarctic or
Nearctic region into a bird's-eye view during the
months that migration is in progress in spring and
autumn, and to be able to see the various species
en route, each following its chosen course, the whole
movement would appear to be in the wildest
confusion. Birds would be seen flying this way
and that, crossing each other's path, or following it
for a certain distance, then quitting it again, passing
each other in opposite streams, or gathering in
certain localities to separate again as soon as they
were crossed, flying from many points of the compass, intricately mixed, yet all orderly following the
chosen path which leads to the desired destination !

As we have already seen, Migration is not a
capricious habit ; we shall soon learn that the
Routes which are followed are almost as ancient as
that habit, and are adhered to in a very persistent
way. Broadly speaking, the majority of migratory
birds visit northern regions in spring and southern
regions in autumn (so far as the Northern Hemisphere species are concerned) ; there is a vast double

exodus from north to south, and from south to north, every year. Any person ignorant of the subject might infer, and reasonably so, that this vast army of pilgrim birds passed to and fro without any regard for order or route ; that each strove as best it could to reach the summer home or the winter quarters. But the very reverse is the case. For birds follow certain routes, and in some cases even modify their habits considerably to do so ; some inland species becoming for the time being littoral ; others, marine species, quitting their coastal habitat to follow a certain fly-line. This persistent attachment to old routes of migration is exemplified in a very remarkable manner by the present geographical distribution of certain species, as I hope presently to show.

The great routes of migration whether over land or sea are closely connected with the configurations of the earth's surface. We may for the sake of convenience divide them into four very marked classes, viz. : Sea Routes, Coast Routes, Mountain Routes, and River or Valley Routes. The first of these highways is the one followed more especially by aquatic birds ; land birds are not known in any part of the world habitually to cross seas for much more than 400 miles, unless there are intermediate stages in the form of islands on which they may rest, if necessary. The two longest routes of migration over the sea which land birds are known to follow, are first in the Atlantic Ocean between the Azores and Portugal, a distance of 900 miles ;

and Madeira, a distance of some 550 miles. There is, however, no evidence to show that the migration of land birds over these islands is of any importance, whilst the few that do occur normally are principally Waders, birds well adapted, as we have seen, for extended flights. The land birds of this group are resident, or only classed as accidental stragglers, and only three species are non-European. That these resident individuals remain unmodified, is however a somewhat convincing proof that many individuals from Europe drift to the islands from time to time and interbreed with them. It is also worthy of remark that with only one exception (the Red-legged Partridge *Caccabis rufa*, which was probably introduced), the fifteen resident European land birds are all species of known migratory habits, a fact which suggests the colonization of these islands by birds that had wandered or been driven out of their usual course during passage. Our second instance is the ocean passage between the continent of North America and the Bermudas. The nearest land is North Carolina, a distance of 700 miles, which, as it is in nearly the same latitude, does not fairly represent the sea passage of possible migrants. North of these islands the nearest land is Cape Sable, about 750 miles distant. This is by far the longest ocean passage habitually taken by any migratory land bird on record; and it is difficult to believe that the flight is ever made from choice, or is customary with some individuals of certain species. The great bulk of the birds that visit these islands on passage are Ducks and other aquatic species; and it seems more than probable that all the land birds which do so are individuals blown out of their normal course along the eastern coasts of North America by violent storms. The passage south (the period when the occurrences are most numerous) is made at a season when storms are remarkably prevalent. Violent gales are almost of weekly occurrence in this region, and seldom fail to bring numbers of migratory birds to the islands. It is suggested that many of these wind-driven migrants may have been carried up by local whirlwinds, and borne out to sea by westerly or north-westerly gales prevailing in the upper atmosphere. The greater number of these birds most probably perish, but a few manage to reach these islands more or less regularly every year, the regular arrivals (such as *Ceryle alcyon, Sciurus novæ boracensis,* and *Dolichonyx oryzivora*) being those whose migrations, either from their altitude or date of progress, are probably most exposed to sudden atmospheric disturbance. Migration, as we have seen, does not generally go on during gales, so that the visits of these birds to the Bermudas seem compulsory rather than voluntary, abnormal rather than regular, and we are perfectly justified in concluding that these islands as well as the Azores are far removed from all ordinary Routes of Migration.

Wherever islands dot the narrow seas between, or fringe the coast-line of continents, migration

over them is both extensive and regular; and it is
also by the aid of islands that such vast areas
of sea are crossed as extend, for instance, between
Japan, the Malay Archipelago, and Australia. In
the same way deserts are crossed by stages from
oasis to oasis. It is also owing to this fact that
oceanic migration is practically unknown among
land birds, and that sea migration is so extensive.
For instance, the Wheatear (*Saxicola œnanthe*) can
and does travel from North Greenland or Spitzbergen to the Equator without having to cross the
sea for more than 300 miles at any point of its
route. From Greenland its fly-line crosses Iceland, the Faroes, the Shetlands, the mainland of
the British Islands, France, the Spanish Peninsula,
the Straits of Gibraltar, and ends in Africa. The
fly-line of birds travelling from Spitzbergen crosses
the Arctic Ocean over several small islands to
Scandinavia, thence either across by way of Fair
Isle and the Orkneys to the British Islands, or
down the continental coast-line to Heligoland, and
onwards to France and the South. There is probably more migration across the Mediterranean
than any other sea in the world. It is the great
divide between the summer and winter quarters of
nearly half the birds of the Palæarctic region; and
yet the passage over it is by no means a promis-
cuous one, but it is made with due regard to what
are evidently very ancient routes. All or nearly
all the migrants from the extreme west of Europe,
including the British Islands, enter Africa by way

of the Straits of Gibraltar, the Balearic Isles, Sardinia, and Sicily; those from Eastern Europe by way of the Greek Archipelago and Candia, the Black Sea and Cyprus. There is also a very considerable amount of migration across the North Sea, which is the route chosen by many birds on their way from Scandinavia to the British Islands, crossing from the Dovrefjeld in Norway to the Shetlands, where the water passage is not more than a couple of hundred miles. Birds either come by this route to our islands, either to winter or to pass still further south; or follow the continental land and cross by way of Heligoland and the Straits of Dover. A very important stream of migration also crosses this sea nearly due east and west, but there is little or no evidence to show that the passage is made across the widest part. It is very significant how few Palæarctic birds extend their migrations to New Zealand, compared with the number that regularly visit Australia, South Africa, or South America. New Zealand is so remarkably isolated that an ocean flight of upwards of 700 miles without a break of any kind is necessary to reach it. Of the great number of migratory birds that leave the east Palæarctic region and China in autumn to winter in the Malay Archipelago, New Guinea, and Australia (across seas thickly studded with islands), scarcely any extend their migration to New Zealand, either wisely declining the long continuous ocean passage, or totally ignorant of that country's existence.

All these facts tend to show that Migration Routes, broadly speaking, are *continuous*, and that it is only under such a condition that Migration ever became a possibility, or even had an existence. Widely disconnected sea routes of migration are as rare as discontinuous areas of distribution, and in many cases not only terminate in emigration, but ultimately in segregation and the establishment of new races. Had there been no means of slowly acquiring the habit in past ages by easy sea passages, or by a continuous land area from north to south or from east to west, had broad expanses of ocean barred the way of retreating birds during periods of climatal change, all the ancestors of our northern avifauna would either have perished or reached southern lands never to return ; or if some did succeed in returning they would have come as emigrants, and migration would have been unknown! Tested by such facts as are here adduced, it would seem that the migration from the Antarctic regions, which in a previous chapter I have suggested took place on a very important scale in past ages, must be an erroneous conclusion. But the difficulty is more imaginary than real. Even at the present day most of the vast ocean space between the now glaciated Antarctic continent and the two great land masses of America and Asia may be so bridged by islands that the continuous water passage does not exceed 500 miles, except to the south of New Zealand (700 miles), and to the south of Africa (1500 miles). From this we may naturally infer

that probably the great highways of Antarctic
migration were Pacific rather than Atlantic, by way
of Australia and South America.   The evidence of
a migration route across the sea south of Africa
is not so abundant or so conclusive, but certainly
exists in the very obvious dispersal of the Swallows
(HIRUNDINIDÆ) from the Antarctic regions, and by
the presence of a few isolated species of CHARADRIIDÆ
in that region.   These latter are few, however, and
unsatisfactory in comparison with the numerous
species (some of them obviously ancestral) left
behind in the Australian and Neotropical regions.
We must also remember that vast and important
geological change has taken place in these Antarctic
latitudes during Secondary and early Tertiary time.
We have direct evidence of land connection between
Australia and New Zealand during the former
epoch, and probably much land has been submerged
(possibly owing to excessive glaciation) between
New Zealand and the Antarctic continent, and even
between South Africa and that region.   The
Ethiopian Swallows are direct evidence of an ancient
Antarctic migration, which, judged by its present
philosophy, must have received land assistance in
its passage south of Africa, although those ancient
stages have long passed away.   At no time, however, judged by the much more exacting botanical
evidence, has the Antarctic region been so closely
bridged with Africa as it has undoubtedly been
with Australia and South America; and this fact
is further emphasized by the notoriously great

powers of wing possessed by Swallows and the great
altitude at which they fly, indicating the longest
possible inter-ocean migratory flights.

From the above facts we may infer that sea
routes of migration are either short routes, or if
prolonged over waters of moderate extent, that
islands invariably assist the passage when taken
under normal circumstances. I also think, as
previously stated, that the altitude of migration
flight is much greater over a sea route than a land
route, for the obvious reason that landmarks are
less numerous and further apart.

The second of these great Migration Highways
is that which follows the Coast-lines of the world.
The contour of the great continents is singularly
favourable to Migration Flight, not only in general
direction, but in its continuity. Coast-lines divide
themselves very naturally into six great groups.
Firstly, we have the East Atlantic coast-line,
stretching in almost unbroken continuity from the
North Cape in Scandinavia to the Cape of Good
Hope in South Africa, a distance which may be
roughly estimated at some 10,000 miles. Secondly,
we have the West Atlantic coast-line extending
from Grinnell Land and Greenland in one majestic
and unbroken course down to Cape Horn in
Patagonia, and covering, we will say, 14,000 miles.
Thirdly, we have the East Pacific coast-line reaching
from Point Barrow in Alaska continuously to Cape
Horn, and which may be roughly estimated at
12,000 miles. Fourthly, we have the West Pacific

H

coast-line, by far the most broken and interrupted of all, reaching from Cape Serdze in north-east Siberia to Tasmania, which may be roughly estimated at some 10,000 miles. Fifthly, we have the West Indian Ocean coast-line, extending from Suez to the Cape of Good Hope, a distance of some 6000 miles; and sixthly, the East Indian Ocean coast-line, reaching from the head of the Persian Gulf to Tasmania, a distance of, say 10,000 miles.

From these facts it may readily be seen that birds which follow coast-lines have practically a continuous road, easy to follow, stretching from one hemisphere to the other, and long enough to include the limits of all but the few very widest migratory flights. Birds do not, however, strictly confine their flight to all the indentations of a coast-line. Were they to do so, the length of their journey would be enormously increased. It may be laid down as a pretty general rule that during actual migration flight along a coast, all the bays are avoided in which the boundary headlands are visible by the species of most powerful wing; that only the narrow inlets are crossed by the less powerful winged; and only the weakest fliers of all follow the winding course of the land. Promontories are also crossed to a very great extent. That this is a fact is easily proved by the vast number of birds that cross a headland, compared with the small number that may be observed in deep indentations of the coast; always excepting, of course, the normal migration into any area, which often enters by such channels. Bad weather occasionally keeps all birds close to the coast-line, and thus modifies their direct line of flight. The coast migration of Eastern America, from what evidence I can glean on the subject, appears to cross from Florida to the West Indies, rather than follow the continental coast of the Gulf of Mexico to South America. I might give many instances of birds on their way south from the Eastern States to South America that cross the West Indies and Trinidad. Another very interesting fact connected with coast migration, is the remarkable way in which certain birds suddenly alter their course by leaving a coast and starting directly across the sea. These birds are supposed to be following old migration routes— ancient coast-lines now submerged beneath the waves. The known great attachment of birds to their fly-lines makes this explanation feasible, and it is still further confirmed by direct geological evidence. There is some evidence, for instance, that birds follow an ancient coast-line once reaching from Spurn Point in Yorkshire to Denmark or Holland, in the fact that several species are known to migrate along the East coast of England up to this point, and then to strike across the sea, seldom or never being observed further north: the Knot (*Tringa canutus*), the Bar-tailed Godwit (*Limosa rufa*), and the Gray Plover (*Charadrius helveticus*), may be mentioned as examples. Another ancient coast-line followed by the Knot, the Asiatic Golden Plover (*Charadrius fulvus*), and the eastern form of the Bar-tailed Godwit (*Limosa rufa uropygialis*), appears to lie between New Caledonia and New Zealand, part of which, in the form of Norfolk Island, still remains above the surface of the ocean. The fly-lines of *Totanus incanus* also appear to suggest a submerged route across the Pacific from Alaska to Polynesia. One more instance, and that perhaps the most interesting, must be given. The geographical distribution of the eastern form of the Orange-legged Hobby (*Falco amurensis*) has long been a puzzle to naturalists. This bird breeds in Eastern Siberia, Mongolia, and North China, and winters in India and South-east Africa; although how it reaches the latter country, and for what reason, has never been satisfactorily determined. The only possible explanation is, that this bird follows an ancient route across the Indian Ocean, much of which has become submerged, although sufficient is left not only to guide the bird on its ocean pilgrimage, but to indicate the position of the sunken land, in the form of the *Maldive Islands*, the *Chagos Archipelago*, the Seychelles, Amurante Island, and the *Saya de Malha Banks* (the names in italics probably show the route followed). The individuals of the Common Hoopoe (*Upupa epops*), that winter in Madagascar, also most probably follow this ancient fly-line, as that bird is otherwise unknown in Africa south of the Equator. That at no very remote age there were considerable land masses along this route (Lemuria), is not only proved by the present conditions of the ocean-bed, but by the presence of many Indian avian types in Madagascar, not only of genera but of species, some of the latter being hardly distinguishable. That it was also at no very remote period a regular route for migrants by way of the Indian Peninsula, is equally certain, as the present migrations of this Falcon undoubtedly demonstrate. The occurrence of the east Palæarctic Cuckoo (*Cuculus himalayanus*) in Madagascar also suggests a passage across this ocean; and the presence of *Cuculus gularis* and *Cuculus capensis* in South-east Africa appears to me to indicate an Emigration by the same route. Future research may yet show that *Cuculus himalayanus* is as regular in its visits to Madagascar as *Falco amurensis*. This route is also, I believe, one of the fly-lines of the Curlew Sandpiper (*Tringa subarquata*) to South Africa, a fact confirmed by that bird only occurring on passage in Madagascar (*fide* M. Pollen). This route of Emigration may also possibly explain, among various other instances, the presence of *Rhynchæa capensis* in Madagascar and continental Africa; and of *Glareola ocularis* in Madagascar, whose nearest ally is the Oriental *Glareola orientalis*; more especially so as it appears even now occasionally to visit the outlying islands (Mauritius, *fide* Grandidier). This route may yet be shown to be a regular fly-line of other east Palæarctic birds, especially of such that are gifted with great powers of flight; for it must be remembered that the widest water-stages are quite 600 miles across. It is probably owing to that

fact that most of the migration by this route has
ceased.

We have only instanced the most important
coast routes of migration, but there are innumerable
others in all parts of the world, both of seas and
large lakes, which are followed by various species
during passage. Witness the vast number of
migrants that pass the coast-lines of the British
Islands, or follow the shores of the Bothnian Gulf,
the Baltic, the White, Mediterranean, Black, and
Caspian Seas in the western Palæarctic region; the
shores of Lake Baikal, and the Great Lakes of North
America. It would require a chapter to particu-
larize them all. The coast routes are more especially
the migration highways of great numbers of
Waders (CHARADRIIDÆ) and Ducks (ANATIDÆ);
the abundance of food is probably the chief reason
for this choice. As soon as the breeding season
is over these birds begin to congregate on or off
the nearest coasts, and gradually move south along
them, many species remaining on them all the
winter in the far south, and passing north again
by exactly the same route. That a vast number
of land birds, especially Passeres, follow coast-lines,
is also equally certain: but these must take such
routes primarily for the sake of an unfailing guide
which trends in the precise direction they wish to
go, rather than from any partiality for littoral
haunts. Coast routes, then, are the recognized
Highways of Migration, followed by hosts of birds
throughout the world. Great numbers of these

birds come from considerable distances inland,
following local routes or bye-ways to join the great
trunk road which all then follow in common
according to their specific time.

On the other hand, however, equally large
numbers of species, and even of individuals, whose
habitat is in the interior of mighty continents, are
too far removed from such coastal highways, either
to be aware of their very existence, or to render a
cross flight impracticable, and these birds have
naturally chosen other configurations of the earth's
surface to serve as great trunk roads for their
annual migrations. Unquestionably the most important of these are the great river valleys which
sere the earth's surface from north to south in so
many countries visited by migratory birds. To a
very great extent all the small tributaries which
drain the haunts of these migrants are followed
until the main valley is reached and the trunk
fly-line joined. On all the great continents there
are river valleys known to be crowded with migrants
passing along them in spring and autumn. In
Europe the most important River Routes are as
follows. Firstly, the valleys of the Petchora, north
and south Dvina, and Onega, which lead to the
various upper waters of the Volga, the Don, and
the Dnieper, along which all the migratory land
birds of Russia journey south to the Black and
Caspian Basins, and thence (by the coast routes
of those great inland seas) to East Persia, Asia
Minor, and Egypt. Secondly, the Vistula, the

Oder, the Elbe, and the Rhine, which connect with
the Danube system, which leads on to Turkey,
Greece, and the East Mediterranean sea routes,
and along which many of the migrants from the
East to the British Islands and from Scandinavia
journey to and fro.   In Africa, which is practically
a continuation of the direct routes from Europe,
the most important river highway is the Nile
Valley, which receives an incredible number of
migrants from Russia, Asia Minor, and Syria—
most of them species that winter far to the south
in that continent.   The Niger Valleys, but on a
much less important scale, drain some of the
migration which spreads across North-west Africa;
but in the west of the continent coast routes predominate.   In Asia the great river valleys are
singularly well situated for migration, and the
number of birds that pass along them on flight
is past all belief.   The migrations of almost every
species of migratory east Palæarctic birds may be
traced along them.   The three great northern
river-systems are those of the Obb, the Yenesay,
and the Lena.   Not only do the main valleys of
these vast waterways favour a direct flight towards
the winter quarters of birds due south of them,
but their endless south-westerly trending feeder
valleys favour the migration of those birds breeding in the east Palæarctic region, and wintering in
the Ethiopian region.   By these cross valley routes
many species whose eastern range in summer extends as far as the Yenesay (which is on the

meridian of Calcutta), journey to winter quarters in
Africa. Some of these cross valley fly-lines are little
short of marvellous. The Great Snipe (*Scolopax
major*) manages partly by their aid to reach
Central Siberia from South Africa, its only known
winter quarters. The Little Gull (*Larus minutus*)
crosses Siberia from end to end by means of these
water-ways, breeding on the shores of the Sea of
Ochotsk, and wintering in the Caspian Basin!
That the latter bird must keep to the water *en
route* is imperative owing to its aquatic habits.
Other great river routes of migration are the valley
of the Amoor, which is principally used by birds
travelling between East Siberia and China, Mongolia and India; the Hoangho and the Yangtse,
which drain Mantchooria and North China, and
feed Burma and India, *vid* the Brahmapootra and
Ganges. Mr. F. W. Styan (*Ibis*, 1890, p. 317)
records no less than ninety-seven species of birds
which regularly pass along the Yangtse on passage,
many on their way to the far north, from the
Siamese Peninsula and Burma, many to cross the
Yellow Sea and breed in Japan. There appears
to be a very considerable stream of migration enter
North-east India down the valley of the Brahmapootra. One very direct instance is furnished by
the Pintail Snipe (*Scolopax stenura*). This bird
breeds in East Siberia, and winters in India, Burma,
the Siamese Peninsula, and the Malay Archipelago.
Its only fly-line into India is apparently down
the Brahmapootra Valley, since the bird is quite

unknown in the north-west of that country.   It
swarms in the adjoining valley of the Yangtse,
which is a continuation of its fly-line to the northern
summer haunts.   Again, it is apparently by this
route that many rare Eastern stragglers to India
enter that empire and follow the Ganges.   The
Crested Teal (*Anas falcata*) and the Baikal Teal
(*Anas formosa*) may be mentioned as instances,
both of which species are common winter visitors to
the Yangtse.   Other important Asiatic river routes
are along the Ganges, the Indus, and the Oxus,
forming the great valley passages into India from
the north-west; and the Tigris and Euphrates,
which assist the passage of birds breeding in
Europe and wintering in Asia.   Instances of the
latter, however, are rare.   The two most interesting
are perhaps the Rose-coloured Pastor (*Pastor roseus*)
and the Black-headed Bunting (*Emberiza melanocephala*), which visit South Europe as far west as
Italy in summer, and are only known to winter in
India.   The great water-ways of North America,
stretching as they do directly and almost continuously from Alaska to Mexico, are also routes
of migration of the highest importance, seeing that
coast routes can only be followed by a very small
percentage of the birds visiting an area in which
the two coasts are nearly 3000 miles apart.   To a
very great extent migration from the North-west
follows the Mackenzie River, and the various water-
ways that lead to the largest lake system on the
earth's surface ; and doubtless a good deal of coast

migration passes along the shores of these vast
inland sheets of water. There is much evidence
to show that the noble St. Lawrence is a fly-line
of many northern Nearctic species; whilst further
south the stream of migration is carried along the
various valleys of the Mississippi, Missouri, and
Ohio. These latter to a great extent carry off the
migrants of the United States between the Rockies
and the Alleghany Mountains. The migrants
from the east of these latter mountains appear to
follow the numerous streams that lead to the
Atlantic coast-line; whilst those from the west of
the Rockies follow similar streams to the Pacific
coast.

Our knowledge of the ornithology of South
America is so meagre in its details that we have
next to no data on which to form an opinion as
to the value of river valleys as highways of migration. There cannot, however, be much doubt that
such favourably-placed valleys as those of the
Parana, the Paraguay, and the Uruguay, trending
nearly due north and south for 1500 miles, have a
very important influence on migration in the wide
districts which they affect.

That migration follows these great inland valley
routes is abundantly proved by what has actually
been observed in them. Wherever competent
observers have noted the seasonal movements of
birds along them, the facts are essentially the
same. Down all these great valleys Migration ebbs
and flows in no uncertain trickling stream, but in

mighty torrents which testify each recurring season
to their vital importance as Highways of pilgrim
birds.   The river fly-lines equal in length those
of the longest coast routes; and although in our
present state of knowledge we cannot trace the
absolute route taken by more than a few individual
species, we are enabled from our general information respecting their geographical distribution
broadly to determine the general course of a very
great number of others.   River valleys are exceptionally favourable migration routes.   The great
variety of species following them can obtain abundance of food either on their waters, on their banks,
or amidst the rich vegetation which clothes the
slopes above the stream.   Thus, we find that birds
of all kinds follow the course of rivers; land birds
and water birds, the insect- or seed-eating Passeres,
the swamp-loving Waders, the aquatic Ducks,
even the oceanic Gull or Tern—all are equally
favoured.   They also enable birds to reach their
breeding-grounds in the Arctic regions at the
earliest possible moment, which is of the greatest
importance in a land where summer, if hot and
brilliant, is remarkably short.

It remains for us now to notice Mountain
Routes.   Although these are perhaps followed least
frequently of all, we have a considerable amount of
evidence to prove that they are not only widely
used, but of very great importance.   As we have
already seen, mountain chains in many cases act as
landmarks and guides to the migrating birds that

follow them, just as coast-lines and valleys do.
They also enable birds to make certain well-recognized and easily-remembered entrances to countries
they pass on passage, or visit during summer or
winter. We have direct evidence that the lower
slopes of mountains are direct highways of migration. Even in England the Downs are a noted
path for migrants; and a great many species may
be traced along the mountain chains of our islands
during the season of their passage. Birds that
belong to a mountain or upland habitat are the
most addicted to these routes. The Ring Ouzel
(*Merula torquata*), various species of Chat (*Saxicola*
and *Pratincola*), the Dotterel (*Eudromias morinellus*),
and a few Waders, are all decided mountain fol-
lowers. We have also the direct testimony of the
most accurate field ornithologists to prove that
flocks of birds on passage may frequently be seen
above mountain ranges, following the chain. Some
of the most interesting instances of migration may
be witnessed at the great mountain passes, birds
journeying through lofty defiles with as much
appreciation of their usefulness as human travellers.
Great numbers of birds pass the Pyrenees, the
Caucasus, and the Alps on migration. One of the
most famous passes for migrants in the Pyrenees
is the " Jaisquivel," another the " Palomeras de
Eshalar"; whilst the " Albula " and " Bermina "
passes into the Adda Valley and Lake Como, in
the Alps, are others. The Himalayan passes are
also great routes of migrants. Dr. Scully, an

ornithologist of wide Indian experience, and for
some time stationed at Gilgit in the North-west
Himalayas, informs me of the wonderful amount
of migration in and out of India of Palæarctic
birds through the passes of this wild upland region
—a stream of migration which is continued along
the valley of the Indus, as previously noted. The
Pamir Plateaux, otherwise known as "the roof of
the world," in Central Asia, is another important
route of migration, as the late Dr. Severtzow's
observations abundantly prove. In North Africa,
the Atlas mountains are followed by many migrants
on their way from Morocco and Algeria to Europe,
viâ Sardinia, Corsica, and Sicily. Our information
respecting mountain routes in the New World is
not very great; but there can be little doubt that
they are important, more especially as the ranges
there are almost parallel to the coast-lines. As
guides they must prove of inestimable service.

From the above remarks it will readily be seen
that the several great Routes of Migration are of a
very varied character; and when in addition to them
we take into account the numberless local routes,
of which it would not be possible to name more
than a tithe, we can form some idea, if only of the
slightest kind, of the complicated nature of these
avian fly-lines. It must also be remembered that
probably very few birds keep exclusively to one
or other of these routes, but make use of all, or
at least several of them, during their seasonal
flights. A bird in its journey from South Africa

to the tundras of Northern Europe or Asia probably
follows all in turn.   It has an experience of Valley
Routes in the Nile, the Don, the Volga, and the
Petchora; of Sea Routes across the Mediterranean
by way of Cyprus or Candia, and the Greek
Archipelago; of Coast Routes along the shores of
the Black Sea; and probably of Mountain Routes
in the Caucasus—each and all of which long ages
of accumulated experience have taught that species
to make the fullest use.   Some of these routes are
much more direct than others; many are excessively circuitous, and vividly illustrate the gradual
way in which a species has spread from a centre
of dispersal, turned this way and that by endless
conflicting influences in the unceasing struggle
for place and for life.   Here a desert stopped the
way, and a fly-line to avoid it had slowly to be
learnt; there some other dominant and vigorous
species already held the ground, causing a *détour*
or even a retreat; here a mountain pass led to
new areas of dispersal, and fields for Emigration;
there conditions of life peremptorily forbade settlement or increase—all this and more is indelibly
stamped upon the present fly-lines of every species,
had we only perceptive power enough to decipher
it.   For, depend upon it, these tortuous Routes of
Migration are the hieroglyphics which record the
Line of Emigration followed by species in past
ages, and unquestionably demonstrate the only
feasible way in which the road has been learnt!

CHAPTER V.

THE subjects of the present chapter are so closely,
I may say inseparably, connected with Migration
that it becomes absolutely necessary to include
them if we desire to make the history and philosophy
of our subject even reasonably complete. We
have had, and shall continue to have, occasion to
allude casually to the Emigration of Birds; we will
therefore devote the present chapter to its discussion in greater detail. There is not a little popular

confusion between the two words Emigration and Migration; by many persons they are regarded as synonymous expressions, and applied indiscriminately to these very distinct avian movements. When the term Emigration is applied to birds, it is intended to express a colonizing movement, a journey with no return, or a spasmodic or gradual extension of geographical area. By the term Migration, a regular passage between two districts or regions is implied. Emigration is either fitful and irregular, or very gradual if constant; Migration is both regular, constant, and seasonal.

The present universal distribution of Birds over the earth's surface can only be accounted for in one of two ways. Either we must admit that every bird was created in the area which it now occupies, or that birds have emigrated in endless directions from centres of dispersal. The former explanation demands the acceptance of the theory of Special Creation, a theory that all the teachings of modern science have utterly exploded, and proved to be as illogical as it is false. The latter explanation, the theory of Evolution, of Descent with Modification, which implies that birds have sprung from common ancestors, is completely in harmony with the facts that are presented to us, not only in the present distribution of animal life, but with the vast changes that our globe has suffered in past ages.

The most important causes of emigration, and those which have probably had the most influence on this means of dispersal, are the great climatal

I

changes that we have already dwelt upon at some
length in an earlier chapter. The two last Glacial
Epochs (at the South and North Pole respectively)
were vast incentives to emigration, causing it to be
undertaken on a scale, so far as the class Aves is
concerned, never equalled before or since those
periods. And not only were the emigrants dispersed from these desolated Polar centres, but their
influx in lower latitudes must have had such a
disturbing influence on avian life in those latitudes,
as to lead, through more severe conditions of life
(owing to competition with invading species), to
much emigration amongst southern forms as well.
The next great cause of emigration is the vast
and almost universal amount of glaciation which
has taken place on every continent during periods
of high orbital eccentricity. "In the Alps," says
Wallace, "the Pyrenees, in the British Isles and
Scandinavia, in Spain and the Atlas, in the Caucasus and the Himalayas, in Eastern North America
and West of the Rocky Mountains, in the Andes,
in the mountains of Brazil, in South Africa, and in
New Zealand, huge moraines, and other unmistak-
able ice-marks, attest the universal descent of the
snow-line for several thousand feet below its present
level." That such ice action produced much
climatal change and direct banishment of organic
life is indisputable; and that the influx of birds from
hill districts to plains and valleys necessitated emigration on a wide scale among upland and lowland
species alike, can scarcely be doubted.

More local, but none the less certain, causes of
emigration may be found in the great numerical
increase of species, rendering the dispersal of the
surplus population into new regions indispensable
and imperative. This may happen in two ways.
Either a vast wave of surplus population may
suddenly spread out from the congested districts—
an irruption, or even series of irruptions, within a
comparatively short period of time, flowing across
wide areas until gradually spent; or, under favourable conditions, a species may slowly extend its
range from a comparatively small centre until it
ultimately covers an enormous area.   As an example
of Irruptic Emigration we have the intensely interesting emigrations of Pallas's Sand Grouse (*Syrrhaptes
paradoxus*) from Central Asia, which have from
time to time occurred with startling suddenness.
This species for the past fifty years or so has
evidently been in a very restless and disturbed
state, and from time to time great waves of emigrants have been thrown out apparently to relieve
a congested area of distribution.   Pallas's Sand
Grouse normally is an inhabitant of the vast plains
or steppes that stretch continuously from north-east
Turkestan and South Siberia to Mongolia.   In the
north it is a migratory bird, and the winter range
extends into North China in the east and the
Kirghiz Steppes north of the Aral Sea in the west.
Until 1859 this species was practically unknown to
western ornithologists, although Russian naturalists
had met with it from time to time in its far eastern

habitat. In that year, however, the first signs of the coming irruptions broke into Europe, and examples of this Sand Grouse were obtained in Poland, Jutland, Holland, and the British Islands. It is interesting to note that the evident direction of this wave of emigration followed a north-westerly course from the Kirghiz Steppes, almost exactly corresponding to the normal north-easterly route. Four years later (in 1863) a much more important irruption took place, this time consisting probably of thousands of individuals, and very much the same route was followed; although, as might be expected in such a great rush of individuals, the wave spread wider and further, extending to Italy and the Pyrenees in the south, to Scandinavia and Archangel in the north, and throughout the British Isles to the Faroes. That these birds were attempting to found new colonies is proved by the fact that many of them endeavoured to breed in places that were best adapted to their requirements. In 1888, another and even more important wave of emigrating Sand Grouse spread over Western Europe, the particulars of which will still be fresh in the mind of the reader. This invasion was undoubtedly the most successful of all; and so well did the birds appear to be established that in our islands a special Act of Parliament was passed for their protection. There can be little doubt, however, that Western civilization will be too powerful a check on their colonizing efforts, and that each irruptic wave having the misfortune to flow westwards into Europe is doomed inevitably to destruction. The Rose-coloured Pastor (*Pastor roseus*) is another species that has evidently permanently increased its western range by very similar means; and even now vast flocks occasionally wander into new districts, like irruptions from congested areas seeking vent for their superabundant life. The observant ornithologist may often remark the occasional and often very extensive irruptions of much commoner birds, species less likely to arrest universal attention, that take place, and this unusual abundance is probably the result of a general exodus of surplus population from some overcrowded district. The irruptic emigrations of such remarkable birds as Sand Grouse or Pastors are noticed at once, whilst those of commoner species are apt to be overlooked, or their importance under-estimated, or even entirely ignored. The vast flights of Common Jays (*Garrulus glandarius*), for instance, that were noticed passing Heligoland in the autumn of 1882, for three days in succession, was probably an irruptic emigration of surplus population from a congested district. Similar irruptic waves of Goldcrests (*Regulus cristatus*) are occasionally remarked. Such instances as the above are more or less exceptional events at the present day, a period, as I have already remarked, of long-continued stability, primarily due to low eccentricity of the earth's orbit; but they enable us to form some slight idea of what Emigration must have been during epochs of great disturbance.

Instances of Chronic Emigration are not only as interesting, but even more numerous, and appeal all the more forcibly to us, because they are either actually in progress around us, or have only ceased during historic time. The evidence that many species of birds have quite recently extended their range, or are even in the act of doing so, is above the faintest suspicion of doubt, and in no small number of cases amounts to absolute proof. The Arctic Willow Wren (*Phylloscopus borealis*) at one time bred in North-east Siberia, and wintered in Burma and the Malay Archipelago, as might naturally be inferred from the locality of its summer quarters. But a slow and gradual emigration set in westwards across Siberia and Europe, and now this species actually visits Finmark in summer, but returns along the old routes of gradual dispersal to the ancient trunk fly-lines of the far East, which it follows to the accustomed winter home! The Siberian Pipit (*Anthus gustavi*) has emigrated gradually from the East in a precisely similar manner, and now its summer range is known to extend at least as far west as the Petchora Valley, in Russia, although its winter quarters are still confined to South-eastern Asia. The Rustic Bunting (*Emberiza rustica*) has extended its range even as far west as Scandinavia in summer, but returns to India and China to winter. Precisely the same kind of emigration has been taking place among West Palæarctic birds. The individual Willow Wrens (*Phylloscopus trochilus*) and Sedge Warblers

(*Acrocephalus phragmitis*), that are now known to migrate as far eastwards in spring as the valley of the Yenesay, return to Africa to winter! The Little Gull (*Larus minutus*) sends out pioneers as far as the Sea of Ochotsk, which return to the extreme south-west of Asia and to Africa to winter with the rest of the species. The Arctic Tern (*Sterna arctica*) is only found during winter in the Atlantic Ocean region, but in summer the range has been so far extended, that a great many individuals spread across Siberia to Behring Sea on the one hand, and across Arctic America to that sea on the other, where they breed in some abundance. Now the most astonishing part of this apparently anomalous distribution, is the fact that these various species go so far to winter quarters, when equally suitable regions might be reached without, in some cases, requiring a journey of more than a fourth of the distance. It seems little short of marvellous that the Rustic Buntings, for instance, breeding in Scandinavia, should return to India and China, and decline to accompany the vast number of European birds that migrate south to winter in North and West Africa, in whose company they have absolutely lived all the summer; or that the Sedge Warblers breeding in the Yenesay Valley should come west again to Africa, parting company with the myriads of Siberian birds, their neighbours of the summer, leaving them to journey down that great migration highway to India, whilst they laboriously push on to Africa, more than double

the distance! But the fact is, these little emigrant
birds only know the way to that winter home, which
has been their winter home from the remotest times,
perhaps as long as their species has had existence;
and they follow routes towards it along which their
emigrations have extended. Their present fly-lines
of migration then are inseparably connected with
the direction of their past emigration, and indicate
unerringly the road the colonists have followed from
the central area of distribution, in opening out new
and ultimately wide tracts of country for their
surplus population.

But although we are able to trace with exactness
the routes of recent emigration, the more ancient
tracks that many species followed from common
centres of dispersal have long been utterly obliter-
ated. The present dispersal of obviously allied
species, however, enables us to trace some of them
almost with equal precision. The least used routes
of Migration are the ones that best indicate the
direction of ancient Emigration. They are routes,
once easily and extensively followed, which from
physical causes (especially submergence) have either
been discarded altogether, or only followed by a
few species. Most of these old routes would be
entirely lost to us were it not for these lingering
migrants across them, or the very obvious near
alliance of the species along the route, where it still
remains partially continuous, or the more distant
yet equally certain relationship of forms at either
end of that broken route.

I have already alluded to what I believe to be two
very interesting ancient routes of migration, which
in still more remote ages were obviously routes of
emigration. One of these extended between India
and South Africa across the Indian Ocean; the
other as surely extended between Eastern Asia
and New Zealand, by way of New Caledonia and
Norfolk Island. The Orange-legged Hobby (*Falco
amurensis*) is one of the last surviving instances
of a fly-line across the Indian Ocean; but the
number of species isolated in South Africa, allied
to Indian species, testify to its ancient importance,
and mark the route, the only feasible route of their
emigrations, between these two countries. The only
other land connection between India and Africa is
by way of Arabia; but if we were to assume such
to have been the route (and there is not the slightest
evidence in its favour), we are confronted with the
difficulty of transporting thoroughly tropical and
south temperate species into northern zones, and
then isolating them in South Africa without leaving
a solitary trace or relic of their dispersal in the
intervening country. Besides, some of the most
interesting species, decidedly Oriental in type, are
isolated in Madagascar, or on the Inter-Indian
islands. Thus, in the Seychelles we find isolated
species of such thoroughly Oriental genera as
Copsychus and *Hypsipetes;* in Mauritius and Rod-
riguez species of *Palæornis*. It needs then no great
stretch of imagination to recall those past ages,
probably towards the close of the Tertiary Period,

when the Indian Ocean was studded with vast island groups between India and South Africa, forming fly-lines of migration between Asia and Ethiopia. Nor does it require any great perception to note how the emigrations of certain dominant species, compelled to increase their area in any suitable direction, by over-population or other equally potent causes, extended from island to island, until the two great countries were bridged by an Avian chain, of which many of the links still exist in sedentary birds isolated in South Africa, and of at least a few migratory birds that still continue to cross by a route, most of which has long disappeared beneath the waves!

Again, the Knot (*Tringa canutus*) and the Asiatic Golden Plover (*Charadrius fulvus*) are two of the last surviving instances of the important route of migration that once ebbed and flowed across the Pacific between Asia, New Zealand, and probably Antarctic land. Not only is this route still pointed out by the few migrants that continue to follow it, but its importance as a still more ancient route of emigration is confirmed in a singularly interesting manner by the present geographical distribution of various species of Ouzels. Emigration among the important group (Turdinæ) of which these birds form a considerable portion, has taken place on a very large scale. That the group is of Polar origin there can be little doubt, and its emigrations have spread far and wide over almost every portion of the earth's surface; not only apparently in irruptic waves, but in steady chronic tides. From south to north, from New Zealand to Japan, stranded races, so closely allied in many cases as to be almost conspecific, on almost every important island or group of islands, point the direction emigration has taken, and suggestively indicate a much more continuous land surface between the Malay Archipelago, North Australia, and New Zealand than is now the case. Thus Norfolk Island is the home of *Merula poliocephala;* Lord Howe's Island that of *M. vinitincta;* New Caledonia that of *M. xanthopus;* the Loyalty Isles that of *M. mareensis* (Maré Isle), and *M. pritzbueri* (Lifu Isle); the New Hebrides that of *M. albifrons* (Eromanga Isle); the Fiji Islands that of *M. bicolor* (Kandavu Isle), and *M. tempesti* (Taviuni Isle); the Malay Archipelago that of *M. javanica;* Formosa that of *M. albiceps.* Here we have evidence of emigration on a very extensive scale; and testing the antiquity of the movement by the close affinities of the species, we are forced to the conclusion that it occurred at no very remote period. It seems to me that this emigration was not of a chronic character, the result of a gradual increase of population, slowly spreading from one island to another, for all of the species are now sedentary, but rather of an irruptic nature, caused by climatal change in a southern centre of dispersal; or even by the concurrent gradual or even rapid submergence of a continuous route of migration between New Zealand and Japan, during a glacial period, which caused their present isolation on the various points which still remain above the ocean. But whatever the cause may have been is of little or no importance, the facts remain as a convincing proof of emigration on a wide and extensive scale.

There is also some evidence to suggest that this vast emigration of Ouzels, besides taking a direct northern course over the Pacific, also followed a route nearly due east across that ocean by way of the Low Archipelago and chain of islands that extends along the line of the Tropic of Capricorn to South America, where their descendants live and flourish in considerable numbers. Or we can account for the presence of these Neotropical Ouzels by an emigration from an Antarctic continent; one stream of emigrants retreating by way of New Zealand and the Pacific Islands; the other by way of Graham's Land, the South Shetlands, and Patagonia. Various faunal and floral links curiously enough bind New Zealand to South America, and the most obvious direction in which the connection once existed is by the now glaciated Antarctic continent. The fact that the Ouzels are not found in the Nearctic and Ethiopian region is very suggestive of an emigration from Antarctic latitudes, because Africa and North America were by far the most isolated from South Polar land; although they were undoubtedly the most important regions invaded by North Polar species during the Post-Pliocene Glacial epoch. The present distribution of the Snipes (*Scolopax*) appears to denote ancient emigrations by routes which are in many places, if not absolutely identical, still almost the same as those followed by the Ouzels! So plainly are these past emigrations indicated by present geographical distribution, that one can almost venture to prophesy the discovery of new species of Ouzel and Snipe in that tropical chain of islands reaching across the Pacific, thus making the area of distribution of *Merula* and *Scolopax* continuous and complete.

Instances of small but recent avian emigration may even be met with in the British Islands. Such species, for instance, as the Song Thrush (*Turdus musicus*), Missel Thrush (*Turdus viscivorus*), and the Rook (*Corvus frugilegus*) are in a more or less acute state of emigration, gradually extending their area of dispersal as circumstances may arise favouring the increase. In Scotland these birds are gradually following the planting of trees; and I found it to be the invariable experience of competent observers in Skye, that soon after a plantation was formed, birds made their appearance therein which had never been met with in the neighbourhood before. The late Mr. Cameron of Tallisker (Skye) gave me many interesting details bearing upon the emigration of our common resident birds. The Partridge (*Perdix cinerea*) has followed the spread of corn cultivation in Scotland. The House Sparrow (*Passer domesticus*) has emigrated far and wide throughout the civilized world. This bird, according to Lyell, made its first appearance on the Irtish when the Russians commenced to till the soil. About 150 years ago it spread up the Obb, and four years later had emigrated 500 miles still further to the east in this river valley. Thence it pushed onwards through the Yenesay to the Lake Baikal district, and is now abundant throughout Siberia, within these limits, wherever civilization has spread. Many instances might be given where various species have gradually become much more abundant, even in our islands. All this evidence tends to show that Chronic Emigration is far from being absolutely quiescent at the present day; rather must we presume that it is ever ready to break out much more acutely, even with irruptic virulence, whenever a stimulating cause may arise.

In many cases chronic emigration, or even irruptic emigration, may lead to the adoption of migratory habits, if the winters of an invaded district be too severe for constant residence therein. I have just alluded to the Song Thrush as a species in a state of chronic emigration in the British Islands. Even a very marked migration takes place during winter, which would lead us to infer that this species has only recently extended its range so far north. In Scotland this migration is even more pronounced, as the Duke of Argyll has most obligingly informed me. He writes from Inveraray: "At this moment our Song Thrushes have just returned. They almost all leave us for the winter season, although the Blackbird never

does. The Song Thrushes return regularly about
the first week in February, or about this very
date, the 10th. I did not see a single bird all
the winter." That this dispersal into cooler areas
has necessitated the adoption of regular migratory
habits is absolutely certain; and in every known
case of this kind of emigration we find that the
migratory fly-lines follow the direction of that
emigration, either to the point where it commenced,
or to the point where the regular trunk line of
passage leading to the usual winter quarters of
the species may be joined.

From the above series of facts I think it will
be reasonably evident that emigration is not only
very closely associated with migration, but is even
occasionally the means of its initiation. But emigration has also played other important parts in
Avian Philosophy. The vast, wide-reaching influence of emigration on the Evolution of Species
can never be sufficiently estimated. Everywhere
we find evidence to indicate that emigration has
been one of the most powerful aids to segregation.
Without it the number of existing species would
undoubtedly be enormously reduced; whilst the
beautiful and wonderful dispersal of avian life
throughout the world would have presented a
very different aspect. The differentiation of vast
numbers of species can be directly traced to
emigration, often leading to the complete isolation
of numbers of individuals, and bringing them
under the influence of new conditions of existence,

which have stimulated and preserved variations in many different directions.

The great emigration taken by many Polar species, for instance, and caused by a Glacial epoch, often followed different routes from a centre of dispersal, causing a separation of the species into as many groups of individuals. In numbers of cases these several colonies remained isolated from each other sufficient time for the individuals of each to become differentiated from a parent form; in some cases so completely that when circumstances again brought each group together in a common habitat, they had not only lost the power to interbreed and produce fertile offspring, but had even acquired or developed various characteristics peculiar to each group. In this way emigration has directly led to the origin of new species. When the Polar ice banished every bird from the Arctic regions, emigration took place on a vast scale towards Africa, Southern Asia, and temperate and tropical America. The members of each species by no means kept together. Some followed a western course, others an eastern course; some emigrated down the coasts of Europe, others down the coasts of Asia; some down the Pacific coasts of America, others down the Atlantic coasts of that continent; or parties of emigrants were divided by the great mountain chains that stretch from north to south in the Nearctic and Neotropical regions. The result of this emigration and isolation is so indelibly stamped upon the birds of the Palæarctic and Nearctic regions of to-day, that we can trace with absolute certainty not only the route many of those ancient bands of emigrants followed, but the ancestral forms from which they sprung. Probably few if any of the species living in the circumpolar region during Præ-glacial times survive at the present day. Some of them were doubtless exterminated; others became segregated into two or more species. Take, for instance, the great number of Palæarctic birds, that are represented in the Nearctic region by closely allied forms, or that are divided into eastern and western races due to emigration and isolation in past ages; or yet again the eastern and western species of Nearctic birds whose areas of distribution are separated by the Rocky Mountains, the result of diverging routes of emigration during the Glacial Epoch. In fact throughout the Northern Hemisphere the ornithological student is continually discovering fresh evidence of the vast influence of emigration on the origin of avian species in these regions. Many instances might be given in support of these statements, did space permit; many I have already recorded in *Evolution without Natural Selection*, a work to which I would refer any reader sufficiently interested to follow the subject further. Precisely the same vast emigration has taken place from South Polar areas due to glaciation.

Again, many island species of birds owe their origin almost entirely to emigration. From a

K

variety of reasons, as we have already seen, birds
are apt to dispose of surplus population by irruptic
emigration.   Flocks of such birds we have every
reason to believe occasionally wander long distances
from their usual habitat, ready to settle in any
favourable locality they may chance to discover.
That such wandering birds sometimes cross wide
expanses of sea is also certain; and it is to these
nomads that the presence of birds on certain islands
lying out of the track of normal migrants is almost
if not entirely due.   These birds settle in their new
home; and owing to isolation from the rest of
the species, any variations that may arise, excited
primarily by changed conditions of life, are preserved through the absence of interbreeding, and
in the process of time become constant characters.
Thus we find in many islands endemic species of
birds obviously descended from parent forms in
adjoining but isolated areas, and to which they
are more or less closely allied.

Islands located far from routes of migration are
almost entirely populated, so far as birds are concerned, by fortuitous emigration.   There is no
regular influx of individuals, as is constantly taking
place at the two seasons of passage on islands
situated on or near a great route of migration, as
for instance at the Bermudas, or even the British
Islands, with the inevitable result of keeping the
sedentary portion of the avifauna true by interbreeding.   Hence we almost invariably find that
the great proportion of the species are endemic, yet

obviously allied to forms on the nearest mainland. Distance is only of minor importance; for some of the most interesting island avifaunas in the world are located close to the continent which has fortuitously given them birth; whilst many remote islands favourably situated on routes of migration are remarkable for their paucity of endemic species. The Galapagos Islands, for instance, are situated on the Equator in the Pacific Ocean, about 600 miles from the West Coast of South America, in a calm region remarkable for the absence of gales. They are far removed from any present highway of migration, and no evidence exists to show any indications of an ancient route ever having crossed them. No less than thirty-eight out of the fifty-seven species of birds hitherto obtained on these islands are absolutely endemic. All the land birds (thirty-one in number) are peculiar except one species, the wide-ranging Rice Bird (*Dolichonyx oryzivora*); and more than half of these thirty species present such divergence of characters as to be classed in distinct genera. The Bermudas, on the other hand, lying 700 miles from the East Coast of North America, and 100 miles further from continental land than the Galapagos, are near one of the greatest routes of migration in the world, situated in an area where equinoctial storms of great violence and persistency prevail. More than 180 species of birds have been recorded from them, yet not one of these is endemic, and of the ten species that are resident, all are common on the adjoining continent, and individuals are repeatedly arriving to mix and interbreed with the island individuals, thus preventing any possible differentiation which might and undoubtedly would soon occur through their isolation. In this case the effects of any fortuitous emigration have been rapidly eradicated, if ever they appeared, by the constant influx of wandering individuals from the adjoining route of migration. So long as this fly-line continues to be recognized by migratory Nearctic species, the Bermudas cannot possibly acquire any remarkable or specialized endemic avifauna.

Even the British Islands can furnish one or two instances bearing on this interesting subject. These islands are remarkably poor in endemic species, partly owing to their separation from continental Europe being so recent, and partly because they are situated on a great route of migration, which keeps the island individuals of almost every species well mixed with continental individuals, and thus by interbreeding checks any tendency to variation being preserved by isolation. Endemic British species of birds are therefore excessively rare. The Red Grouse (*Lagopus scoticus*) is certainly the most interesting, and may have been the result of a fortuitous emigration of Willow Grouse (*Lagopus albus*) from Scandinavia, or a colony of the latter left isolated on British moors by the submergence of land between the Orkneys and that country. Any way, the Red Grouse owes its specific distinctness to the fact that its continental ally is sedentary in the sense of not crossing the sea. The St. Kilda Wren (*Troglodytes hirtensis*) is another instance. This island form of the Common Wren (*T. parvulus*) succeeds in retaining its distinguishing characteristics, not only because it is sedentary in St. Kilda, but because its island home is a long way removed from the usual fly-lines of any migrant Wrens that cross from the continent to our islands. This endemic race shows how rapidly variation can take place when its greatest check, the facility of interbreeding with the parent form, is removed. On the other hand, such slight local variation as is presented in the British form of the Coal Tit (*Parus ater britannicus*) and the Long-tailed Tit (*Acredula caudata rosea*) are prevented from becoming more specialized or constant by the regular influx of individuals from adjoining continental areas, which visit our islands and interbreed with these local forms. Again, Heligoland (with Sandy Island, a tiny islet of only some 250 acres area) has a wonderful record of no less than 396 species reputed to have been met with on its shores, but does not contain one endemic bird, because it is situated on another important route of migration. The British Islands, being one of the best examples known of recent continental islands, furnish wonderful evidence of the emigration of species. At a period no more remote than the latter part of the Post-Pliocene Glacial Epoch, nearly if not all their area was submerged to a depth of some 2000 feet, only our highest mountains remaining above the sea in scattered rocky islets. This awful devastating submergence banished all or nearly all living things; so that from the period of their subsequent elevation the tide of emigration must have set in towards them in no uncertain stream, to people these islands with their present wealth of plant and animal life! Many bird emigrations to them led to migration, or increased the fly-lines of species; and these routes of migration then slowly formed are followed with amazing persistency down to the present time!

From the above facts we can form some idea of the vast importance of Emigration, not only in the dispersal and segregation of avian life, but in the periodical movements of birds. The great geological, geographical, and astronomical changes in past ages have driven birds to and fro across the earth and ocean; excess of population or invasion of competing races have despatched them from endless centres to every part of the world capable of supporting and nourishing them; whilst the results of their mazy peregrinations are indelibly stamped upon existing species, and much of the direction of these ancient emigrations are indicated by the present dispersal of birds over the earth's surface.

CHAPTER VI.

INTERNAL MIGRATIONS AND LOCAL MOVEMENTS.

IN addition to the prolonged migrations that many birds undertake, there is an immense amount of Internal Migration and Local Movement in progress amongst others at certain seasons of the year. Some of these internal migrations are as regular as the more extended flights, follow certain routes, take place at appointed times. Indeed, so universal are the causes leading to migratory movement of some kind, that the probability is, very few birds indeed can be regarded as thoroughly sedentary. Even in the lowlands of the equatorial regions, where the usual type of migration is unknown among the resident avifaunas, considerable movements take place according to season.

We will deal with the phenomenon of Internal Migrations first. These may be divided very naturally into two great groups, viz. Vertical Migration, or the regular passage of birds from the plains to the hills; and the Northern Flights of various species breeding in the temperate portions of the Southern Hemisphere, and whose order of progression is exactly the reverse of what takes place in the Northern Hemisphere. The amount of vertical migration is enormous, and is most prevalent in hot countries, although there is a very perceptible vertical movement even in such temperate districts as the British Islands and Scandinavia. Perhaps in no other country is vertical migration more pronounced or more widespread than in India. Here great numbers of species retire from the plains to the slopes of the Himalayas to breed, ascending thousands of feet above sea-level, and returning to the lowlands as the cold season approaches. The Common Woodcock (*Scolopax rusticola*) goes at least to an elevation of 10,000 feet to breed in these mountains, and winters on the plains. The movements of these birds are just as regular as those of species whose fly-lines extend for thousands of miles, although the altitude visited varies a good

deal in individual species. The same remarks
apply to North Africa. Many birds there are
regular migrants, coming up from the desert in
spring, where they have been spending the winter
in the various oases, and breeding on the slopes of
the Atlas. Their fly-lines may be traced down
certain valleys, through gorges, and from oasis to
oasis, followed just as unerringly as those of birds
whose migrations extend far across the sea.

I had the pleasure of observing two very interesting instances of this vertical migration whilst
travelling in Algeria, one of which had hitherto
escaped the notice of naturalists. This was the
vertical movement of Tristram's Warbler (*Sylvia
algeriensis*), a species that was originally discovered
by Canon Tristram in the remote oases of the Sahara,
and whose habitat was stated by that naturalist to
be ".only in the southern desert." This region,
however, is but its winter quarters, for in summer
I found it distributed throughout the Djebel Aurés,
from the plateau of Batna, 3500 feet above sea-
level, up to 6000 feet near Oued Taga. The
second species was the gay and lively Bush Chat
(*Pratincola mousseri*), which Canon Tristram found
in increasing numbers as he went south into the
desert during winter; but in summer exactly the
reverse conditions prevail, and I found it equally
common from Batna up to 6000 feet, becoming
less common on the lower and southern slopes of
the Atlas to the oasis of Biskra, which is only 360
feet above sea-level. Again, the Chat (*Saxicola*

seebohmi), a species discovered by Captain Elwes
and myself, is only known to breed at 5500 feet
elevation in the Aurés, and doubtless winters in
the oases of the desert.   The Goldfinch (*Fringilla
carduelis*) is another instance.   I found this bird
breeding in Algeria, where it is a resident, up to
4000 feet, and wintering on the plains.   In fine,
wherever mountains occur, it may be laid down as
an almost universal rule that there is a considerable
amount of migration taking place between their
slopes and the plains.   This rule, however, is by
no means confined to species resident in those
countries; for we find in a great many cases that
some individuals of a migratory species wintering
in southern lands and breeding in the Arctic or
temperate regions, ascend mountains to such
altitudes as render the climatal conditions similar
to those prevailing in the higher latitudes to which
the bulk of the individuals resort.   This is a very
remarkable and interesting fact, of which the following instances may be regarded as typical.   The
Dotterel (*Eudromias morinellus*) breeds on the
tundras of the Arctic regions above the limits of
forest growth, but a few individuals find a similar
climate at high elevation in the Alps, and on the
mountains of Great Britain and Scandinavia.   A
few Tree Pipits (*Anthus arboreus*) breed on the Alps
and the Pyrenees, but the great majority migrate
north in spring.   The Redstart (*Ruticilla phœnicurus*)
winters in North Africa, passes through South
Europe on migration, and breeds throughout Central

and Northern Europe up to the Arctic Circle. A
few individuals, however, ascend the mountains of
South Europe to breed in the pine region. The
Wheatear (*Saxicola œnanthe*) has a very similar
range, but extending much further north; a few
individuals ascend the highest mountains of South
Europe to breed in the pine and birch regions.
The Whinchat (*Pratincola rubetra*) breeds sparingly
on the mountains of South Europe. The Black-
throated Ouzel (*Merula atrigularis*) breeds in Central
Siberia, and winters in Baluchistan, India, and West
Turkestan; but many individuals ascend the Himalayas and the mountains of Turkestan to the pine
region, where they breed.

We also find that in many sedentary species
ranging, say from sub-tropical to north temperate
regions, the individuals in the extreme southern and
warmest limits of the range ascend mountains to
breed, where they find similar climatal conditions
as those individuals dwelling in the more northern
and cooler portions. Hence, one section of the
species is resident, the other portion has acquired
migratory habits, although the journey is vertical
instead of latitudinal. The Hedge Accentor
(*Accentor modularis*) is one of the most familiar
instances occurring to me. This species is practically sedentary, except in the extreme northern
portions of its range (although even here it is said
in some places to be resident), which extends
throughout Europe south of lat. 70° in the west,
and lat. 64° in the east. Everywhere in the extreme

southern limits of its range it retires to mountains
to breed.

Much shorter yet equally interesting movements
in a vertical direction are furnished by a great many
endemic mountain species. Wherever the mountains
are lofty enough to range from a tropic or sub-
tropic zone at their base to a temperate or Arctic
climate at their summit, we have numerous instances
of vertical migration. In summer these birds ascend
into the region that affords them the requisite climate
during the season of reproduction, some of course
going to much greater elevations than others, to
the rhododendron, the pine, or the birch region, as
the case may be, and descending with the approach
of winter to a more genial climatal zone. Many
instances of this kind occur in the Caucasus, the
mountains of Turkestan, the Himalayas, the Andes,
and elsewhere. In the Alps and the Carpathians,
for instance, we find the Alpine Accentor (*Accentor
alpinus*) an endemic species, which visits the highest
summits to breed in the Arctic climate above the
limit of forest growth, and just below the line of
perpetual snow, retiring in winter to the lower
valleys. The vertical migrations of some of the
Rose Finches (*Carpodacus*) are precisely similar,
some of these birds ascending in summer to an
elevation of 10,000 feet, and wintering in the lower
valleys.

From all these facts it will be seen that vertical
migration is very similar to latitudinal migration,
and that it is undertaken for purposes precisely the

same. Probably much of it was initiated in equatorial regions during periods of intense local glaciation; whilst many of these mountain migrants in
more temperate zones may be the last survivors of
the hosts of birds that were driven from Polar zones
by the Post-Pliocene Glacial Epoch; remaining
behind to breed in regions that were once on the
immediate margin of the glaciers, and in their
movements at the present day very clearly indicating the nature and extent of that migration that
prevailed during the acute phases of Polar glaciation
in past ages. Whilst other species, even other
individuals, have gradually extended their northern
flights towards that olden Polar Paradise, these have
remained content with shorter pilgrimages; although,
in every instance, it will have been remarked that
the object attained (a similarity of breeding temperature) is precisely the same.

We now pass to the migration of birds in the
Southern Hemisphere. Unfortunately we have
many obstacles to contend with, and are placed at
considerable disadvantage in our study of bird
migration in this region. In the first place, the
data on which any conclusions may be based are
somewhat meagre, partly owing to the scarcity of
any careful and intelligent observation, and partly
owing to geographical peculiarities, rendering
migratory movements not only few but exceedingly
restricted, in comparison with the vast flights of
northern species. Nevertheless migration in the
Southern Hemisphere, as I hope ultimately to show,

is not only of intense interest in itself, but of vital importance as an indicating demonstration of physical mutations and biological changes as vast and far-reaching in their results as any that the Northern Hemisphere has experienced.

The apparently anomalous fact that very few birds breeding in the Southern Hemisphere during summer in the south are known to migrate north of the Equator to winter during summer in the Northern Hemisphere, is a profoundly important one—a fact which in reality is the key to the phenomenon of migration as it is practised during present time. Broadly speaking, every migratory bird throughout the world leaves a warmer climate or zone to breed in a cooler climate, either by ascending mountains until the altitude furnishes the degree of temperature necessary, or by visiting temperate or Arctic regions where similar conditions prevail. A vast number of species then pass from tropical climates to temperate and Arctic latitudes, probably because this area is much more extensive and suitable than the restricted mountain regions in lower latitudes. In these vast northern areas there is no lack of room, and an abundance of food, resulting in easier conditions of life, and consequent decrease in racial struggle for existence. But, as we have already seen, these Polar regions of avian Paradise are by no means eternal; banishment waits upon the bird world there, in the form of glaciation and the complete reversal of climate at either Pole in the course of equinoctical precession combined with orbital eccentricity. The inevitable deduction from these clearly demonstrable facts is, that whichever Pole is passing through a period of mild climatal conditions and freedom from glaciation, the region round that Pole will be the great breeding area of birds whose migrations are latitudinal. Hence at the present time we have few birds undertaking a latitudinal migration during summer in the Southern Hemisphere, and possibly only one or two breeding sufficiently high in South Polar latitudes to bring them into the Northern Hemisphere to winter during our summer, as by the law that the further north a bird goes to breed the further south it goes to winter, and inversely, by inference, the further south a bird goes to breed, the further north it goes to winter, they undoubtedly would do. The reason for this is, that the now glaciated condition of the Southern Pole has destroyed the once fair Antarctic paradise, the Mecca of migratory birds during the ages the Northern Pole was labouring under its desolate burden of glaciation. At the present day conditions are exactly reversed. The Arctic and north temperate regions are free from glaciation, and furnish suitable breeding-grounds for these migrants, and as a natural consequence we find the progress of migration reversed, and birds come north to breed and go south to winter, as we during historic time have only known them to do, and as appears to us therefore the only normal procedure. Migration then in the Southern Hemisphere, as we now see it, is but a fragment of the vast and regular passage that undoubtedly took place when the Antarctic continent—a region we must remember estimated to be six millions of square miles in extent, or twice the area of Australia—sustained and nourished avian life, and is precisely similar to that migration which took place during the Glacial Epoch in the Northern Hemisphere. Almost the only birds left in this region are those that bred in the lower or temperate zones, and, as in the Northern Hemisphere to-day, these species breeding in the lower zones are remarkable for the comparative shortness of their migration flights. Not only so, but many of the species now breeding on the outskirts of the glaciated southern continent, either penetrate to the Falkland Islands, Tierra del Fuego, the South Shetlands, and possibly Graham Land, or obtain suitable climatal conditions by vertical migration; these latter being species that may have lingered in the Southern Hemisphere long after the great Antarctic breeding-grounds had been closed.

It is a very interesting fact that the now prevailing migrations of birds in the Southern Hemisphere confirm the views above expressed. If these premises are true, we should not expect to find any extension of migration flight into the Northern Hemisphere during winter in the Southern Hemisphere. Neither do we, save in a few exceptional cases. Let us test the truth of our conclusions by a comparative examination of this

Southern Hemisphere migration, with what is taking place in the Northern Hemisphere. In the first place, we must not lose sight of the important fact that no non-glaciated land area of any great extent now exists beyond, say south lat. 55°, which contracts the zone of southern breeding-grounds to a latitude corresponding with that of Edinburgh and the Baltic in the Northern Hemisphere. It may be laid down as a pretty general rule, that birds breeding up to the limits of the temperate zone in the Northern Hemisphere, and wintering below the Equator, are few, and principally birds that visit the most northerly portion of that zone, especially Waders. On the other hand, the birds breeding in the Southern Hemisphere and wintering above the Equator are very few (so far as is at present known), because the south temperate zone does not extend far enough south. It is, however, a profoundly interesting fact that in South America and Australia, where this zone extends the furthest south, we find, as we should expect to find, the most northern Migration Flights, some few species being known to come up north to Brazil and New Guinea. The Patagonian Plover (*Charadrius falklandicus*) visits the Falkland Islands and South Patagonia in September and October to breed, and is known to migrate at least a couple of thousand miles north during the antipodean winter, which is just as important a flight as that of the Kentish Plover (*Ægialophilus cantianus*) from England to North Africa. The Falkland Dotterel (*Eudromias*

L

modestus) breeds in the islands whose name it bears, arriving in September and leaving in April, and its northern migrations extend at least 1500 miles to Uruguay, where it was obtained by Darwin. The allied race of this species (*Eudromias modestus ru`ecola*) visits Tierra del Fuego during summer to breed, and is known to migrate during winter for about 2000 miles north along the South American coasts. In the extreme South of Africa many species of Swallows and certain Cuckoos are all migratory, and leave the comparatively cool climate of that region during winter for haunts extending more or less towards the Equator, although the Flights are not perhaps so long as in South America. The Australian Swallow (*Hirundo frontalis*) breeds in Australia, and migrates north to the Equator to winter in New Guinea. It is an equally suggestive fact that several species of Petrel breeding on the borders of the glaciated Antarctic continent, some of the most southerly breeding of birds, regularly visit the Northern Hemisphere during our summer. So interesting and so vitally important to the views here expressed are these northern migrations, that it will be necessary to note a few of them in detail. One of the best known Petrels that comes north during winter in the Southern Hemisphere is Wilson's Petrel (*Oceanites wilsoni*). This bird is known to breed on Kerguelen Island, one of the few islands that lie on the borders of the Antarctic continent, and may possibly do so on other land even nearer the South Polar region. During winter

it migrates north across the Equator to the northern
coasts of the Indian Ocean; in the Atlantic, as high
as the West Indies, New York, and the British
Islands (where flocks are occasionally observed);
in the Pacific, as high as Peru and Chili, and
possibly much further. From May onwards this
Petrel is one of the commonest birds met with in
the Atlantic by the various Liners that cross from
Europe to the States. Wilson's Petrel arrives at its
Antarctic breeding-places in November, and stays
for a period of about five months until the young
are safely reared, then migrates northwards to enjoy
a second summer in the Northern Hemisphere, but
not to breed. Our second instance is that of the
Sooty Shearwater (*Puffinus griseus*), whose only
known breeding-place at the present time is the
Chatham group in nearly the same latitude as
Kerguelen, but in the South Pacific. It migrates
northwards after the breeding season, and has then
been met with as high in the Northern Hemisphere
as the coasts of Newfoundland, Labrador, and
Greenland, the Faroes, and the British Islands;
whilst in the Pacific it is known to range as high
as California. Our third instance is the Collared
Petrel (*Œstrelata torquata*), a species breeding in
the New Hebrides, 2000 miles south of the Equator,
which also comes north to winter, and has been
obtained off the British coasts so recently as
November 1889. That this species is thoroughly
a Southern Hemisphere one seems proved by the
fact that this latter example *was in moult.* Another

instance is furnished by the Cape Petrel (*Daption capensis*), a species said to breed on South Georgia. Now it is all nonsense to attempt to explain these northern migrations away by suggesting that breeding-grounds of these birds remain yet to be discovered in the Northern Hemisphere. The migration of a Petrel from the Island of Desolation, as Kerguelen is otherwise called, to the British Seas, is no more wonderful than the flight of a Knot from Grinnell Land to South Africa. So far then from being in any way anomalous, these northern migrations are perfectly regular, and just what we ought to find if our views on Migration are correct. It must also be remembered that Petrels, the most southerly breeding of birds, and consequently the most northerly ranging during the antipodean winter, are very similar in appearance to northern species, apt to be overlooked, and are seldom shot at a season when the collecting of sea birds in British waters is forbidden by law. Again, our own Petrels breeding furthest north retire in precisely the same way to far southern latitudes, where they are even less likely to be observed, being so thoroughly oceanic in their habits. The Great Shearwater (*Puffinus major*), for instance, breeds as high as South Greenland, and has been obtained near Cape Horn, although naturalists whose knowledge of Migration Philosophy seems none too extensive, have sought to cover the record with discredit, and to imply an error in identification! The Dusky Shearwater (*Puffinus obscurus*) breeds

on the Bermudas, Bahamas, Madeira, &c., and
wanders south in winter, even as far as Australia
and New Zealand. It was my intention to devote
a chapter entirely to Ocean Migration, but the want
of reliable information has reluctantly compelled
me, for the present at any rate, to remain silent.
Migration Flight, however, seems just as regular
and important among oceanic birds as in more
terrestrial species, and to be governed by the same
laws.

These northern flights of Southern Hemisphere
species may yet be found to be more numerous
when the ornithology of the Neotropical region
especially is better known. At the present time
it is one of the least known regions in the world.
In South Africa and Australia, as we should
naturally expect, the northern migration in the
antipodean autumn is the most restricted, for there
we find the least difference between an extended
temperate zone and the Equator.

It is owing to these circumstances that in
northern latitudes we observe very few migratory
birds during summer from the Southern Hemisphere
from an Antarctic region, giving up their little lives
to idleness and enjoyment, side by side with Northern
Hemisphere species busy bringing up their young
and full of family cares—an anomaly that may be
witnessed everywhere, in more or less frequency,
during the antipodean summer, when our migrants
are away from us, and Southern Hemisphere birds
are breeding. These migratory birds breed only

once in the year, either in the Northern Hemisphere
or in the Southern Hemisphere; and the alleged
instances of certain northern species breeding in
South Africa during their winter sojourn are
entirely unsupported by reliable evidence. There
is one circumstance, however, bearing on this
question, to which I should like to call attention,
with the view of obtaining more definite information. Several species of birds known to breed
in the high north have often been observed in
flocks during summer in that region. Is it fair to
presume in every case, as we are apt to do, that
these flocks are young non-breeding birds, perhaps
born the previous year? May it not be possible
that some of these birds have bred in undiscovered
Antarctic breeding-grounds, and are spending the
period of the southern winter in corresponding
northern latitudes? I think more definite information is required as to these gregarious individuals,
seeing that we have some not altogether untrust-
worthy evidence of such thoroughly Polar species
as Bonaparte's Sandpiper (*Tringa bonaparti*),
breeding on the Falkland Islands; the Eastern
Golden Plover (*Charadrius fulvus*), breeding in
New Caledonia (the only records, June and July,
of this species at Heligoland are very suggestive);
and the Turnstone (*Strepsilas interpres*), breeding
on Lord Howe's Island (young partially fledged
have been captured in this island); whilst the eggs
of the Curlew Sandpiper (*Tringa subarquata*), a
species that goes as far south as Australia to winter,

are entirely unknown. Is it possible that the Knot (*Tringa canutus*) breeds anywhere in the Antarctic regions? We know that this bird passes to and fro between the Polar regions of either hemisphere in vast numbers, and still no breeding-ground has yet been discovered in the Arctic regions in any way proportionate to those numbers. Mr. Hudson in his lately published valuable work, *The Naturalist in La Plata*, remarks the appearance of certain Northern Waders on the Pampas (notably *Limosa hudsonica*), at a season which strongly suggests their having bred in Antarctic latitudes. Again, it is quite possible that many individuals of species that winter in South Africa and breed in Europe, as the Swallow (*Hirundo rustica*), the Willow Wren (*Phylloscopus trochilus*), and the Sedge Warbler (*Acrocephalus phragmitis*), for instance, might visit the Northern Hemisphere in our summer, after having bred in South Africa, but do not attempt to breed again, and be overlooked. No one would suspect such a thing to be taking place, and yet it is not impossible, if not very probable. We should expect to find this state of things prevailing, if at all, in the Northern Hemisphere, on the southern limits of the summer area of dispersal of these species; in Algeria, for instance, where curiously enough all three of the above-named birds are found throughout the year. Probably somewhere in Central Africa these birds may be found all the year round, yet never breeding in those equatorial districts. It is also a curious fact that the Quail

(*Coturnix communis*) is a spring visitor to and breeds in South Africa; and this seems to confirm the view that there is somewhere in Central Africa a Neutral Zone of non-breeding birds of various species, part of which come north to breed in the Palæarctic region, and the other part go south to breed in the temperate portions of the Ethiopian region. As yet there is no evidence whatever to show that the breeding area of the Quail is continuous. The Black-necked Grebe (*Podiceps nigricollis*) is another good instance. From what we have already observed, I strongly suspect that this Neutral Zone will eventually be discovered; it is postulated on this evidence, especially when we bear in mind that there are no localities suitable for the breeding-grounds of decidedly temperate species in equatorial Africa. Brazil and the Malay Archipelago may also contain such Neutral Zones. These, however, are questions connected with the science of Migration that must be left to future research to solve. I allude to them, because it seems to me they suggest a way to great discoveries.

There is one other point connected with migration in the Southern Hemisphere which tends to confirm the views previously expressed on this subject, and that is the isolation of many species of birds in the Southern Hemisphere obviously nearly allied to northern types. This is just what we should expect to find upon the Antarctic regions becoming glaciated. Most if not all of these upland species in the temperate regions of the Southern Hemisphere are birds banished by the South Polar Glacial Epoch; many of them are sedentary on the mountains; others have acquired regular habits of vertical migration, just as we have seen is the case with many species in the Northern Hemisphere.

We thus see that the study of Migration in the Southern Hemisphere is a very important one, for it enables us to test the soundness of the views we have expressed, and tends to confirm them in no uncertain way. When migration has been studied in this part of the world as diligently as in the Northern Hemisphere, and we have consequently the same abundance of material from which to make deductions concerning it, the nature and purpose of this grand and important avian movement in the Antipodes will not be found to differ in any important respect from that prevailing in the opposite hemisphere.

We now pass to the second portion of the subject of the present chapter, the Local Movements of Birds. These local movements are almost if not entirely confined to the season of winter in the temperate zones, and to the dry season in the torrid zone of the earth. They are indulged in not only by endemic species in each of these great regions, but by migrants whilst sojourning in their winter quarters. Probably the only cause of these local movements, wherever they occur, is due to failure of food supply. Although these movements have been little studied by naturalists, especially in

the warmer regions of the world, we have abundant
evidence that they are not only very common, but
strongly marked.   Thus even in the tropics, where
life of all kinds seems perennial, birds wander about
at stated periods in quest of favourite food.   We
have the evidence of naturalists who have noticed
these errant wanderings of species in tropical forests,
that various birds only appear in certain districts
during the flowering or fruiting of certain trees.
This has often been remarked in the case of
various Humming Birds and Parrots : "When the
parasite plants of Guiana," says Waterton, "have
come into full bloom, then is the proper time to
find certain Humming Birds, which you never fall
in with when these parasites are only in leaf.   I
have sought for them whole months without
success, until the blooming of the parasite plant
informed me that I need labour in vain no longer."
Further, it has repeatedly been noticed that many
species of birds in the tropics are distributed over
certain parts of the area of their dispersal according
to season; coming to some districts to breed, and
retiring to others as soon as that duty is completed.
During the hot season in some countries great areas
are so burned and scorched that many birds are
compelled to migrate for some distance to other
areas, where more suitable conditions of existence
are presented.   The distance travelled, the routes
followed, and the exact periods of absence, have
been little recorded; but the broad fact remains
that a movement takes place.   Probably very few

species in any part of the world remain absolutely
stationary throughout the year; everywhere important changes take place, and birds have to adapt
themselves to those changes, which in most cases
involve a temporary removal from one district
to a more or less remote other district. In
countries where vast flights of locusts are continually
wandering to and fro, birds of many species follow
in their wake to prey upon these insects; whilst
in South Africa, Mr. Seebohm observed a most
interesting local movement of certain birds in search
of roasted grasshoppers, destroyed by the great
prairie fires. Large flights of Pratincoles (*Glareola
melanoptera*), and numbers of Ruppell's Lapwings
(*Vanellus melanopterus*), and Birchell's Coursers
(*Cursorius rufus*), follow these fires from one district
to another, to feed on the abundant fare they
provide. The various local movements of the
ubiquitous Rice Bird (*Dolichonyx oryzivora*) of
North America are equally interesting; as are even
these of our own House Sparrow (*Passer domesticus*).
This latter bird is subject to much local movement
during summer and autumn, and wanders far and
wide in flocks in quest of grain. The Lapwing
(*Vanellus cristatus*), the Snipes (*Scolopax*), especially
the Woodcock (*Scolopax rusticola*), and the Sky
Lark (*Alauda arvensis*), may be instanced amongst
numerous others as British species that wander
about in winter, often in considerable numbers, in
quest of food. Birds of the Pigeon tribe
(COLUMBIDÆ) are notorious wanderers; so are the

various berry-eating species, such as the Fieldfare (*Turdus pilaris*) and the Missel-Thrush (*Turdus viscivorus*). All these birds undertake journeys of varying length during winter in quest of food— movements not exactly of a migratory nature, yet sufficiently regular and important to require notice in connection with the usual migration flight of birds. Again, birds of far-extending and regular migration wander about their winter quarters after their long Flight is done, visiting this district and that according to the abundance of food.

It is very difficult to classify these Local Movements, or to determine which are regular migrations and which are not. It would seem that a certain amount of local migration is actually in progress even during mid-winter, as the evidence gathered by the Migration Committee of the British Association, by Gätke on Heligoland, and other observers, is absolutely undeniable. Nor is it confined to the colder regions of the world, for instances are not wanting of Winter Flight in the tropic zones. It is evident that a considerable amount of Winter Flight takes place over the North Sea, to and fro between the Continent and the British Islands, especially among Waders and aquatic birds; and this movement is probably due entirely to failing food supply, or severe weather in either district respectively. A long spell of severe weather sends great flights of birds from one district to another where milder conditions prevail. I have repeatedly observed instances of this winter migration during severe weather. During long-continued snowstorms all our Sky Larks have vanished; frosts of long duration invariably banish the Redwing (*Turdus iliacus*), the Snipes (*Scolopax*), and other ground-feeding species; failure of the berry supply will initiate a local migration of all birds that chiefly depend on it for subsistence. On the other hand, during severe weather flocks of other birds have visited us that seldom or never do so under ordinary circumstances. We shall enter more fully into this in a later chapter.

From the above series of facts we may learn that few birds are really stationary throughout the year; that it is rather the exception for a species to be absolutely sedentary. Again, many if not all young birds are great wanderers, driven from their birth-place by their parents, or deserting it voluntarily as soon as parental care becomes unnecessary. One important result of all this Local Movement is, that it serves to keep individuals well mixed together, and insures that all-necessary cross-breeding, or "mixed marriage," which is so essential to the well-being and even preservation of each and every species.

CHAPTER VII.

NOMADIC MIGRATION.

Nomadic Migration most prevalent in Cold Regions—Resident Birds in the Arctic Regions—The CORVIDÆ as instances of Nomadic Migration—The Pine Grosbeak and the Shore Lark—Snow Buntings —Arctic Grouse—Ducks and Gulls— Nomadic Migration in Antarctic Regions—What Nomadic Migration teaches—Birds of Short Migration Flight most closely allied to Nomadic Migrants—Claim of Nomadic Migrants to Generic Rank—Absence of Representative Forms of Nomadic Migrants in the Southern Hemisphere —Geographical Distribution of the Shore Larks.

THERE is another class of migrants it now becomes necessary to notice, birds whose periodical flights are too important perhaps to be classed with mere local movements, yet too irregular to come within the scope of any migratory movement hitherto described. The species indulging in this peculiar kind of migration are the Nomads of the Avian world, the restless wanderers with no settled or definite winter home. Just as the nomad savage wanders to and fro about his wilderness, pitching his camp here one day, miles away the next, according to his ever-fluctuating supply of the bare necessaries of life, so do these vagrant birds pass the non-breeding season in quest of food. We find the greatest amount of Nomadic Migration prevailing among birds peculiar to the coldest regions of the earth, either on mountains or in high northern latitudes—species whose supply of food is not curtailed by any decrease of temperature, able to live throughout the long Arctic winter wherever the snow does not absolutely cover the various substances on which they live.

These nomadic migrants very forcibly show that Want of Food was one of the great initiating causes of regular migration in autumn, just as High Temperature was probably the great initiating cause in spring. So long as food can be obtained, most birds show great reluctance to adopt a migratory habit in autumn; therefore throughout the Arctic regions, except, perhaps in the Polar zone, some birds may be found all the winter through wherever food can be obtained. No insectivorous birds are known to winter above the isothermal line of pre-vailing snow and frost at that season, but many species of birds whose food consists of buds, twigs, seeds, and berries; or that subsist on any carrion or refuse; or that prey upon these other birds themselves, habitually remain near this snow-clad and frost-bound area, moving about just as the food supply may fluctuate, sometimes wandering south during a spell of unfavourable weather, but hasten-ing north again as easier climatal conditions recur.

Of course the birds that habitually winter in the Arctic regions are comparatively few, for the simple reason that the great majority of species that visit this area are either exclusively insectivorous, or subsist on a variety of animal food that cannot be obtained during winter. Some of the most interesting instances of nomadic migration are presented by the various species of CORVIDÆ that frequent the Arctic regions. The Raven (*Corvus corax*) keeps to the extreme north as long as food can be found, and wherever a village or settlement furnishes any regular supply of refuse, will brave all the rigours of an Arctic winter with impunity. The Siberian Jay (*Perisoreus infaustus*), one of the most warmly clad of all Arctic birds, keeps to the northern forests, wandering about to the more open and cultivated districts during winter or unusually severe weather, returning again as soon as sufficient food can be found. The Magpie (*Pica caudata*) and the Nutcracker (*Nucifraga caryocatactes*) dodge about their Arctic haunts throughout the winter, wandering hither and thither, and frequenting the villages and the post-roads to pick up a living, retiring to their more accustomed haunts as soon as the weather permits. None of these birds are migratory in the strict sense of the term ; neither, however, are they by any means stationary ; they are birds at the mercy of circumstances—wanderers and nomads, either becoming gregarious at the approach of winter, or remaining solitary or in pairs. The Pine Grosbeak (*Pinicola enucleator*) is another thorough nomadic migrant. It lives in summer in the more open forest districts of the Arctic regions ; in winter it gathers into flocks like other

Finches. These flocks of Grosbeaks then wander
far and wide according to circumstances, and though
their southern migrations occasionally extend as
far as the British Islands, France, and Hungary,
there is no southern locality to which they regularly
resort in winter, sometimes appearing in one, sometimes in another, just as the weather may affect
their movements. Their stay, even when they do
happen to visit these lower latitudes, is short and
fleeting, and a northern migration commences as
early as climatal conditions permit. The Shore
Lark (*Otocoris alpestris*) is another Arctic nomad
with no regular winter quarters, wandering about
at that season spending its time wherever it can
find food. Sometimes it wanders to the British
Islands, and there is evidence to show that its visits
are gradually becoming more regular, and the bird
itself more numerous. This may indicate some
change in higher latitudes affecting this species,
necessitating more regular passage, a fact from
which we may learn how readily a Nomadic Migration may develop into regular Passage if the causes
are intensified.

We have many of these Nomadic Migrants that
pay us uncertain and irregular visits, appearing in
the British Islands during some winters, and never
being seen again perhaps for years in the same
abundance. Some of these nomads are more regular
in their appearance than others. Scarcely a winter
passes, for instance, that does not bring Snow
Buntings (*Emberiza nivalis*) in varying numbers;

but the winters are very few during which we see the Pine Grosbeak. The Snow Bunting thoroughly deserves its name ; it is perhaps the very first Passerine bird to penetrate into the higher Arctic regions with the return of spring, long before the snow has melted, or winter relinquished its iron grasp. As soon as the northern peasants begin to throw manure on their snow-clad fields, the Snow Buntings, previously hovering on the very edge of the snow-wreath, make their appearance ; and although later snow-storms may banish them, again and again they return until winter is finally con-quered, and the south wind brings sudden summer on its wings. As a rule, the endemic forest birds of the Arctic regions travel the shortest distances south, and very few of these birds have ever visited our islands. They are nomadic enough in their northern forests, but are rarely, if ever, driven from them to any great distance. The three species of northern Grouse, the Capercailzie (*Tetrao urogallus*), the Black Grouse (*Tetrao tetrix*), and the Hazel Grouse (*Tetrao bonasia*), inhabiting forest districts, wander about more or less during winter, but rarely if ever undertake any migration even of a nomadic character. The Willow Grouse (*Lagopus albus*), however, is a nomadic migrant in very cold areas, in summer frequenting the moors, like its congener the Red Grouse (*Lagopus scoticus*), of the British Islands ; but in winter, when the tundras are several feet deep in snow, a migration is undertaken to the nearest forests, where the birds subsist on buds,

shoots, and pine-needles. I have known the Red
Grouse make similar nomadic movements during
heavy snowstorms, wandering miles from the moors,
and even visiting farmyards and towns. Again,
many of the Crossbills (*Loxia*) are typical nomadic
migrants, having no regular season of passage or
route of flight, wandering up and down as it were
on the fringe of winter, now north, now south, in
sympathy with each recurring change. Various
species of Arctic Ducks (ANATIDÆ) and Gulls
(LARIDÆ) are also good examples of this nomadic
migration. Many of these birds never wander
much south of open water during winter, unless
compelled to do so by violent gales, ice-floes, and
snowstorms. Steller's Eider (*Somateria stelleri*) and
the King Eider (*Somateria spectabilis*) breed on the
coasts of the Arctic Ocean, and the adult birds
rarely come further south than where they can find
open water during the long Polar winter; young
birds, however, as is customary, wander further
south, but never more than in a nomadic fashion.
The Ivory Gull (*Pagophila eburnea*), the Snow Bird
of the Arctic navigator, is another of the few
resident species in the Polar zone, and its nomadic
migrations are short, irregular, and uncertain. It
lives amongst the eternal ice, and its omnivorous
tastes enable it to pick up a sustenance in regions
where most other creatures would inevitably perish.
The Little Auk (*Mergulus alle*) is very similar in
its movements. Its grand head-quarters are the
dreary coasts of Spitzbergen (although it breeds in

suitable places throughout the extreme North Atlantic), where it is a partial resident, but numbers wander about a good deal in winter; and at that season it occasionally appears off the British coasts as a nomadic migrant.

The phenomenon of Nomadic Migration has been little studied in the Antarctic region; but there can be little doubt that it exists among the few peculiar species of birds that dwell on the borders of the glaciated South Polar lands. We cannot expect to find it so frequent or so marked a movement as in the Arctic regions, for obvious reasons, still the movement would well repay careful study, for I am of opinion that it throws much light on the origin of the more regular and extended migrations of birds.

The facts to be derived from a study of Nomadic Migration are of great value in assisting us to understand the origin of regular migration. Many of these nomadic migrants are probably descendants of those species that dwelt on the fringe of the glacial ice during the Post-Pliocene Glacial Epoch, the birds that wandered least from their devastated Polar haunts. There is not a single trace of their migrations ever having been Inter-Polar. In their wandering movements during present time they are profoundly interesting examples of migration in its incipient stage; they illustrate very vividly the rudimentary portion of that grand migration flight which now takes place almost from Pole to Pole. That some few birds remained in as high latitudes

as possible, even during the periods of greatest
glacial intensity, seems very probable; just as round
the margin of the glaciated South Pole we now find
a few birds—last relics, we are compelled to regard
them, of that rich and abundant Antarctic avifauna
that dwelt round the Southern Pole in long past
Eocene ages, and which was scattered north probably by a Post-Eocene Glacial Epoch. From
Nomadic Migration we can gradually trace the
movement into short but Regular Passage; and
thence through every gradation to those long
extended flights which we justly look upon with
admiration and with wonder. These nomadic
migrations represent incipient migration in the past,
which never developed in these species or their
ancestors to any greater extent than what we now
witness; but, as we have already seen in a vast
number, in the great majority of others it gradually
became a function of the highest importance.

It is also worthy of remark, that the species performing the shortest regular migrations are closely
allied to these nomadic migrants. There are few,
if any, wide-ranging migrants among the Crows
(CORVIDÆ), or Finches (FRINGILLIDÆ), Ducks (ANA-
TIDÆ), or Auks (ALCIDÆ), or Gulls (LARIDÆ), unless
belonging to different genera, which indicates more
distant relationship. This fact, so far as it goes, is
one very good reason for placing such thoroughly
Arctic and isolated species in genera to themselves,
in spite of the remonstrances of some naturalists,
who always seem to think that a genus cannot be

a natural division unless it contains a good round
number of species. If the species are few, that is
no reason for lumping them into one or two genera,
at the cost of concealing some of their most interesting features. Thus the Shore Larks fairly claim
distinction from all other Larks, under the generic
name of *Otocoris ;* the Grosbeaks from all other
Finches under that of *Pinicola ;* the Eider Ducks
under that of *Somateria ;* the Little Auk under that
of *Mergulus ;* the Gulls under *Pagophila, Glaucus,*
and *Rhodostethia.* In my opinion, if a genus
illustrates or implies any important fact, either
geographical or biological, it ought to be retained,
even if the species it contains are few.

It might also be remarked of these nomadic
migrants, that none of them are represented by
closely allied forms in the Southern Hemisphere,
and many are exclusively Arctic, a fact of great
significance, indicating very restricted migratory
movements through all Avian time, and probably
confined close to the limits of glaciation. The
geographical distribution of the Shore Larks (*Otocoris*) or the Waxwings (*Ampelis*) illustrate this. I
will select the former genus because it seems the
most anomalous, and yet after all it is not abnormal
in any respect. All the Shore Larks are nomadic
migrants, or actually sedentary, and number six
more or less clearly defined species or races, all
obviously very closely allied. Probably they formed
one circumpolar species previous to the Post-Pliocene
Glacial Epoch. Driven south by the advancing

glaciers, they were ultimately isolated in several
colonies, some in the Nearctic region, some in the
Palæarctic region. What led to their ultimate
specific distinction we need not stay to inquire,
because that does not bear on our present object.
What I want to show is, that various species of
Shore Lark are left isolated along the line which
probably marked the small limits of their emigration
during this Glacial Epoch. Two species occur across
Central Asia from Palestine to China (*Otocoris peni-
cillata*,[1] and *Otocoris longirostris*) ; one is an inhabitant of North Africa and Arabia (*Otocoris
bilopha*) ; two inhabit the New World, one in the
Northern United States (*Otocoris occidentalis*), and
one in the Southern States, Mexico, and Central
America (*Otocoris chrysolæma*). The Common
Shore Lark (*Otocoris alpestris*) has succeeded in
becoming circumpolar once more, as its common
ancestor was in Præ-glacial times, and occupies the
belt of country above its more southern represent-
atives. If the limits of glaciation, as demonstrated
by actual geological evidence, are followed round
the Northern Hemisphere, we shall find these races
of Shore Lark left on its extreme margin, nomadic
migrants then as now!

[1] Four examples of this species have within the past year or
so been obtained in Bosnia.

CHAPTER VIII.

THE Migration of birds is beset with dangers and full of perils. It would scarcely be possible to over-estimate the mortality among birds of passage directly due to migration. One very significant proof of this great mortality is presented in the fact that of the immense numbers of birds flying south or west in autumn, only a very small per-centage come north or east again in spring! Most

people have remarked the great gatherings of
Swallows, Martins, and Swifts, just previous to
migration in autumn, yet where do we see such
similar multitudes in spring? The majority of
these birds are young ones, neither so strong of
wing nor so robust of frame as their parents, and
it is among these that the highest mortality is
reached. The death-rate of a large town standing
at say fifty or sixty per 1000, creates something
like a panic among its human inhabitants; but
there can be no doubt whatever that the death-rate
among birds on migration reaches ten times that
amount per 1000, and during exceptional circumstances very much more! From the moment that
a migrant bird sets out on its journey it is exposed
to quite a new set of dangers, whilst many other
ordinary perils of its existence are very much intensified. From one end of its fly-line to the other
successive dangers surround it, and enemies of
every kind have to be eluded. Migration then,
instead of being a pleasant path in the wake of
retreating summer or in the van of advancing
spring, is the most fatal undertaking in the life of
migrant birds, and few there be that survive it.

The Perils of Migration may be divided into three
important classes, viz. those arising from Fatigue,
due to the mechanical portion of migration flight;
those arising from the Natural Enemies of each
species; and those arising from Blunders and
Fatalities on the way. Probably the first class of
perils is the most fatal one; a journey with little

rest by the way of even a couple of thousand miles, is a great strain on the endurance, especially of small Passerine birds; whilst a sea flight of, say 300 miles, with no opportunity for rest of any kind, and in many cases not even the chance of snatching a mouthful of food *en route*, must tax these tiny migrants to such an extent that only the strongest survive the journey. Of the countless thousands of birds that perish during migration, by far the greater number probably succumb at sea. Many instances are on record of great numbers of drowned migratory birds being washed ashore, especially after stormy weather. Some of these tired migrants save themselves by getting a chance rest on some passing ship, but the majority, especially when flying by night, quietly drop into the remorseless sea and perish! I have seen the Nightingale (*Erithacus luscinia*) rest on a steamer in mid-Mediterranean, as well as the Turtle Dove (*Turtur auritus*) and the Quail (*Coturnix communis*), all being remarkably tame, the former perching on the soldiers lying asleep on deck. These birds were crossing from North Africa to Europe towards the end of April, and only remained a short time with us, probably because we were steaming nearly due south, and therefore taking them out of their way. Had there been no friendly vessel within sight on which to rest, none of these birds probably would ever have reached the European coast.

The number of birds met with in the Atlantic, at varying distances from the British coasts, is very interesting, and gives us some idea of this cause of mortality amongst migrants. The following extract from the *Report on the Migration of Birds*, during 1880, communicated by Mr. Robert Gray, will be read with interest :—

" The ship *Rutland* of Greenock, Captain Roy. When about 400 miles on this side of Newfoundland, during continued heavy gales from the east, Captain Roy observed numbers of birds taking refuge, on the 20th September. He had had head winds all the way home to England, and birds more or less numerous round the ship till the *25th October.* When he was 400—500 miles from Ireland, a violent storm arose, and blew prodigious flocks of birds before it. The deck and rigging were covered. Many died, and many were killed and used as food. The survivors, after staying a few days, were carried off by the force of the wind. Captain Roy observed one Robin, lots of Linnets, Snipe, Thrushes, Wagtails, etc. Heavy rains accompanied the storm." Mr. Gray continues : " My brother-in-law, on his way to Boston in one of the Cunard steamers, saw a Jackdaw and a Starling come on board on Oct. 23rd, during a gale from the east, when 550 miles from the Irish coast, easterly winds having prevailed for several days. On Oct. 24th, 850 miles from land, one Starling perched for a few minutes. A small bird, like a Linnet, hovered about the rigging but did not perch. One Water Rail was captured and detained ten [? two] days, 1200 miles from land, and two

Sandpipers. . . . When about 1080 miles from
Ireland two Crossbills flew on board. Both were
captured."

The arrival of a flock of migrants on the coast
in a more or less exhausted state is some indication
of the loss at sea. Only the strongest have sur-
vived the stormy passage, and even many of these
are so tired and worn out as to allow themselves to
be taken in the hand. On several occasions I have
had the good fortune to witness an arrival of Goldcrests (*Regulus cristatus*), on the east coast of England. Before sunrise on the chilly late October
mornings, I have seen the stunted thorn-bushes on
the dunes or links for miles along the coast
swarming with these tiny creatures—the smallest
migrant in the entire Palæarctic region. Some
have been much more exhausted than others; some
have actually rocked to and fro with weakness as
they sat upon the twigs; but the more robust ones
were feeding eagerly, and some even indulged in
song !

Birds seldom commence an extended migration
flight, especially across the sea, until the weather is
favourable; but a sudden change of wind, a gale,
or a rain- or snow-storm, or a heavy shower of hail,
frequently overtakes them, and beats and chills the
very life out of all but the strongest. Even aquatic
birds, able to drop on the water and rest whenever
they may feel the inclination to do so, are often
overtaken by gales, and blown long distances out of
their way, even far inland. Small wonder then that

a gale or a storm proves fatal to migrating land
birds, many of them frail of form and constitu-
tionally weak and feeble.   We can therefore readily
understand why it is that migrants decline to make
any very extended flights across the sea, except the
most fleet-winged and robust.   On every known
sea-route of migration it has been remarked that
birds, especially small terrestrial species of comparatively feeble flight, choose, invariably, the easiest
way across, where choice is possible.   Some of these
sea passages are, of course, wider than others, but
every available assistance is made the most use of.
More than a hundred years ago, a brother of Gilbert
White had remarked that the Swallows in passing
the Straits of Gibraltar took the narrowest route,
flying over the bay in a south-westerly direction,
and crossing opposite to Tangier.   Even after a
long flight is safely accomplished, the resultant
fatigue renders the poor little migrants utterly incapable of escape from the numerous enemies that
lurk along the route, or follow like the most remorseless and bloodthirsty of pirates in their
wake.

The greatest enemies of birds on migration are
the various Hawks and Falcons and Owls that are
always hovering in close proximity to the defenceless
moving throng. The large Falcons follow migrating
Ducks for enormous distances, preying on the helpless birds at will; the various smaller Hawks fare
right royally on the Warblers, Thrushes, Finches,
and such-like Passerine species.   Each district

through which their fly-lines extend has its own
Raptorial birds waiting to levy tribute as the
migrants hurry by. Hawks have also been observed
to pay long visits to islands lying on routes of
migration, to prey upon the passing birds, and
even to hunt for small migrants in the rays of the
lanterns. Many of these birds of prey themselves
are migratory, and are careful to make their own
seasons of passage correspond with those of their
unfortunate victims. The arrival of the Peregrine
(*Falco peregrinus*), for instance, in the Arctic
regions, takes place simultaneously with that of
the Ducks ; the Hobby (*Falco subbuteo*) does not
appear until late in spring, when the woods are full
of small birds ; but the Merlin (*Falco æsalon*)
follows the very first few venturesome migrants
northwards.

We must not overlook another very fertile source
of danger to migrating birds, and that is the fatal
attraction of lighthouses and light-ships. The
brilliant light of the various beacons that stud the
coasts of the civilized world, along the direct route
of countless migrants, allure birds from their direct
course, especially during cloudy weather, and great
numbers kill themselves as they fly dazed and
bewildered against the glass. It has been universally remarked by light-keepers, that birds strike
most frequently on dark cloudy nights, with fog,
haze, or rain. Instances of birds striking on clear
nights are exceedingly rare. Again, light-ships are
more attractive to birds than lighthouses on islands

or headlands, probably because the former are the *only* visible objects on which to rest or approach, no land being near them: fixed white lights are also more deadly than the revolving or coloured lights. It is also interesting to know that fog-horns prevent birds from striking, for it has been observed that whenever one of these warning engines has been erected near a light where birds previously struck in abundance, the striking has almost ceased. Sea birds rarely strike these lights, although instances are on record where they have done so, Stormy Petrels having been known to cause light-keepers much annoyance by fluttering against the lantern and ejecting oil on the glass. Many intensely interesting scenes may be witnessed from the balcony of a lighthouse or the deck of a light-ship, favourably situated on or near a great highway of migration. Odd birds are continually coming in contact with the lights throughout the migration season; but now and then they appear in wonderful numbers, just as some great flight of birds has been suddenly deflected from its course by a fog or bank of clouds hiding the moon or stars, and throwing the surface of the earth or sea into such shadow that all knowledge of locality is temporarily lost. Here is the report of Mr. Littlewood, the keeper of the *Galloper* Light-vessel, moored fifty miles off the mouth of the Thames, made on the night of October 6th, 1882: "Larks, Starlings, Mountain Sparrows [Bramblings], Titmice, Common Wrens, Redbreasts, Chaffinches, and Plover were picked up on the deck, and it is calculated from 500 to 600 struck the rigging and fell overboard; a large proportion of these were Larks. Thousands of birds were flying round the lantern from 11.30 P.M. to 4.45 A.M., their white breasts, as they dashed to and fro in the circle of light, having the appearance of a heavy fall of snow. This was repeated on the 8th and 12th, and on the night of the 13th, 160 were picked up on deck, including Larks, Starlings, Thrushes, and two Redbreasts; it was thought 1000 struck, and went overboard into the sea." Telegraph-wires are generally placed too low to be in the usual path of migrants, but instances are on record where birds have killed themselves by flying against them on passage. Great numbers of birds also lose their lives every year by flying unwittingly into nets spread along the coast of the Wash. The coast here is a well-recognized highway of migrants coming to our islands from the East, and birds of many species are annually taken in them. Curiously enough, the Woodcock (*Scolopax rusticola*) is rarely or never caught thus. He migrates by night like so many of the rest, but flies high until well over the land, when he drops almost perpendicularly into the most likely cover. Many Woodcocks are foolish enough, however, to commit suicide against lighthouses and vessels; and they have been known to break glass three-eighths of an inch in thickness by the force of contact—evidence, so far as it goes, of the great velocity of migration flight of some species. At the Bell Rock Lighthouse, for instance,

the keeper reports on the nights of October 31st
to November 1st, 1883 : " What we think were
Woodcocks struck with great force. Birds continued flying within the influence of the rays of
light till the first streak of day, continually striking
hard all ·night, and falling into the sea. Although
we cannot be sure, we think there was a great number
of Woodcocks struck, and fell into the sea." The
force with which some birds strike is terrific. A
Little Grebe (*Podiceps minor*) struck the lantern
of the Hasbro Light-vessel, off the Norfolk Coast,
at 11 P.M. on the night of March 30th, 1883,
with such force as to split the bird from the neck
along the entire length of the body ! The keeper
of the Longstone Lighthouse reports on the night
of November 10th, 1885 : " One of the large Snipe
struck S.E. side of lantern at 9.45 P.M., and was
almost smashed to pieces." Again, as recorded in
a Göttland newspaper : "A curious circumstance
happened at the Färö Lighthouse on the 20th
October. About 8.30 in the evening a sharp report
was heard by the man on watch. He immediately
went up to the lantern to ascertain the cause, when
he found two of the panes of glass broken into
small pieces, as well as three Long-tailed Ducks
(*Harelda glacialis*) lying dead inside. There were
also three lamp-glasses broken and a third pane of
glass cracked in many places. On the ground
below lay nine more birds of the same sort. The
Ducks had come from a northerly direction. The
flight must have been at a remarkable speed, as the

N

quarter-inch glass was smashed into many small pieces." At the Hasbro Light-vessel from Oct. 10th 1883, to January 3rd 1884, no less than 162 Sky-Larks, 73 Starlings, 23 Chaffinches, 60 Larks and Chaffinches, 1 Brambling, 3 Fieldfares, 3 Thrushes, 1 Blackbird, 1 Redwing, 2 Goldcrests, 1 Hooded Crow, 1 Rook, 1 Kingfisher, 1 Tame Pigeon, 1 Lapwing, 3 Ringed Plovers, 4 Gray Plovers, 1 Oystercatcher, 1 Curlew, 1 Whimbrel, 2 Woodcocks, and 28 Stormy Petrels, "besides a large tub and bucket-full various," struck and were killed against the lantern, or were picked up on board! At the Tuskar Rock Lighthouse, off Wexford, 1200 birds were counted as killed in a single night, whilst hundreds more fell into the sea! I might continue giving hundreds of instances of fatalities against Lights, but the limits of my space forbid.

The last most important peril of the road is the danger of losing it. Strange as the fact may seem to the uninitiated, great numbers of birds blunder every year, take the wrong direction at some important point of the journey, and find themselves eventually in countries thousands of miles to the east or west of their proper destination! It is difficult to form any estimate of the number of these little erring migrants every year, but there can be no question that it is very considerable. The young birds are the greatest blunderers, the birds that have practically no knowledge whatever of the road, and have to depend entirely on the guidance of older birds. That this is the case is abundantly proved by the fact that nearly all the birds that accidentally wander to the British Islands from more or less remote countries are birds of the year.

The list of British birds abounds with the names of wandering species, not only from Eastern Europe and Siberia, but from Africa, and even from America.[1] These represent individuals that from some cause or another have blundered at the cross-roads, been blown far out to sea, or joined the wrong stream of migration, and journeying in its company have found themselves at last in Western Europe, in the British Islands, instead of in the Malay Archipelago, in India, in Africa, or Mexico. Without including nomadic migrants, whose movements are so erratic that there is nothing extraordinary or abnormal in their occasional visits to our islands, we have a list composed of birds essentially migratory, that may well excite our wonder when the details of each occurrence is studied. Take, for instance, the appearance of White's Thrush (*Geocichla varia*), the Siberian Ground Thrush (*Geocichla sibirica*), and the Black-throated Ouzel (*Merula atrigularis*), all birds regularly breeding in Central and Eastern Asia, and retiring to winter quarters in India, China, and the Malay Archipelago. Or even more re-

---

[1] It should be remarked that stray birds from the east and north generally appear in autumn, as they do also from America; but birds from Africa and from the south as a rule arrive in spring—individuals that have overshot the limits of their normal spring migration.

markable still, the occurrence of the frail and tiny Yellow-browed Willow Wren (*Phylloscopus super-ciliosus*), a species breeding no nearer to us than the pine forests in the valley of the Yenesay, and wintering in India and China, and which to reach us must have flown more than 3000 miles across Asia and Europe, due west, instead of 3000 miles south into India, with an experience of a sea flight (across the German Ocean), which is novel to the migration of this species!   How persistently these erring individuals must have stuck to their route, every mile of it never traversed before, and only guided by the hosts of migrants following their normal course ! No less remarkable are the visits of such birds as the Desert Chat (*Saxicola deserti*), and the eastern race of the Black-throated Chat (*Saxicola stapazina*), from Turkestan. Or yet again, the visit of the Needle-tailed Swift (*Chætura caudacuta*), from localities at least 3000 miles to the east, a bird whose regular winter quarters are at the Antipodes ! Then we have the visits of the Nearctic Yellow-billed · Cuckoo (*Coccyzus ameri-canus*), representing a direct flight of 4000 miles or more, with a possible rest at the Bermudas and the Azores by the way ! Or the by no means less wonderful appearance of the American Bittern (*Botaurus lentiginosus*), a bird that was actually first made known to science from an example shot in Dorset ninety years ago !

Common British species are just as likely to blunder on the way as others. The Common

Snipe (*Scolopax gallinago*) has wandered to the Bermudas; the Common Heron (*Ardea cinerea*) has visited Greenland more than once; the Corn Crake (*Crex pratensis*) has strayed as far as the Atlantic States of America, and even to New Zealand! Again, the Redwing (*Turdus iliacus*) has been known to visit Greenland, the Woodcock (*Scolopax rusticola*) New Jersey and Virginia! Lost birds from one region are continually turning up in other regions throughout the world; it is the rule rather than the exception. As the avifauna of each country becomes more closely studied, these instances of lost birds will undoubtedly be found to be more numerous than is at present suspected. Of the thousands of birds that annually lose their way on migration, but very few are ever captured or recorded. It is only when these wandering individuals enter a district bristling with enthusiastic ornithologists and sportsmen that they are liable to be captured; and even then not one in ten is probably observed.

Whilst on the subject of lost birds our attention naturally turns towards Heligoland, the place of all others where watch has been most strictly kept for their appearance. The publication last year of Gätke's long-expected and long-delayed volume, *Die Vogelwarte Helgoland*, enables us to form some idea of the number of birds that lose themselves on passage; but after all, the list of birds occurring on the much-vaunted island is, to say the least, disappointing. Perhaps we expected too much. Again, it is most exasperating to find records based on evidence of the most flimsy kind, and the occurrence of such a species as *Turdus pallens*, for instance, resting on the identification of a bird-stuffer, who only saw and did not obtain the bird! Or, worse still, *Phylloscopus fuscatus* is admitted to be a bird of Heligoland, because an example was supposed to have been seen! The evidence is only too apparent that every means has been used to swell this list to its greatest possible length, just as certain British naturalists seek to increase the British list on worthless testimony, a method that may suit the collector and the dealer, but ill becomes the man of science. I can safely say I have never met with a list, claiming to be authoritative, in which so many species have been inserted on the most trumpery evidence, during the entire course of my experience. It is an insult to ornithological common sense, and in many ways gives the much-maligned *Ibis List of British Birds* a long start and a good beating! Bad as this Heligoland list undoubtedly is, it contains, however, certain items of profound interest, illustrating the world-wide wanderings of birds. Some of the most wonderful abnormal bird-flights are those taken by migratory Nearctic Thrushes, such as the visit of *Turdus swainsoni* and *Turdus migratorius* in October 1869 and 1874 respectively. No wonder the latter example was found dead after an ocean flight of so many thousands of miles! None the less remarkable are the occurrences of *Mimus carolinensis, Dendrœca virens,* and *Dolichonyx oryzivora,* the latter the wide-ranging Rice Bird, which, as we have already seen, wanders to the Bermudas and the Galapagos. It is also very interesting to remark how so many species of Phylloscopi (of which our Willow Wren, *Phylloscopus trochilus,* is a typical species) lose their way on migration, and travel West from Central Asia, instead of South to India and elsewhere. Individuals of the various species of East Palæarctic Buntings (*Emberiza*) blunder almost equally as much.

All this is very interesting, but we expected much more from the way Heligoland has been lauded for the past twenty years. Now what are the real facts of the case ? The list of birds obtained on Heligoland includes no more than forty species that have not hitherto been observed in the British Islands, although there can be no doubt whatever that many of them have visited us. As a set-off against this, however, the list of British birds contains no less than fifty wanderers (not nomadic migrants) that have not yet been noticed on Heligoland; and we may reasonably conclude that they have not visited the island under any ordinary circumstances, or they would either have been captured or at least identified by such sharp eyes as can determine *Phylloscopus fuscatus* from every other Willow Wren (at least twenty-five species), even when alive and at liberty !

The following table will serve to demonstrate these facts :—

British Species not observed hitherto on Heligoland.	Heligoland Species not observed hitherto in British Islands.
1. Vultur fulvus	1. Circus pallidus
2. „ percnopterus	2. Lanius meridionalis
3. Elanoides furcatus	3. „ isabellinus
4. Astur atricapillus	4. Muscicapa albicollis
5. Surnia hudsonia	5. Turdus swainsoni
6. Saxicola isabellina	6. „ fuscatus
7. Geocichla sibirica	7. „ ruficollis
8. Regulus calendula	8. „ migratorius
9. Tichodroma muraria	9. Mimus carolinensis
10. Loxia leucoptera	10. Erithacus philomela
11. Zonotrichia albicollis	11. Ruticilla mesoleuca
12. Fringilla canaria	12. Phylloscopus bonellii
13. Agelæus phœniceus	13. „ tristis
14. Scolecophagus ferrugineus	14. „ proregulus
15. Sturnella magna	15. „ coronatus
16. Chelidon bicolor	16. „ borealis
17. Progne purpurea	17. „ viridanus
18. Chætura caudacuta	18. „ nitidus
19. Caprimulgus ruficollis	19. Hypolais polyglotta
20. Ceryle alcyon	20. „ elaica
21. Cuculus glandarius	21. „ caligata
22. Coccyzus americanus	22. Acrocephalus agricola
23. Coccyzus erythrophthalmus	23. Locustella certhiola
24. Coracias indicus	24. „ fluviatilis
25. Turtur orientalis	25. Dendrœca virens
26. Ectopistes migratorius	26. Saxicola morio
27. Botaurus lentiginosus	27. Motacilla citreola
28. Butorides virescens	28. „ melanocephala
29. Ægialitis vocifera	29. „ viridis
30. Vanellus gregarius	30. Alauda pispoletta
31. Totanus solitarius	31. „ tartarica
32. „ flavipes	32. Emberiza aureola
33. Ereunetes griseus	33. „ cæsia
34. Tringa bonaparti	34. „ cia
35. „ minutilla	35. „ pityornis
36. Sterna leucoptera	36. „ pyrrhuloides
37. „ hybrida	37. Dolichonyx oryzivora
38. „ fuliginosa	38. Carpodacus roseus
39. Larus melanocephalus	39. Hirundo rufula [1]
40. Colymbus adamsi	40. Larus affinis
41. Puffinus obscurus	
42. Oceanites wilsoni	[1] Recorded also as British.
43. Œstrelata torquata	
44. „ hæsitata	
45. Bulweria columbina	Note.—Possibly Emberiza cioides casta-
46. Daption capensis	neiceps should be included among the
47. Bernicla glaucogaster	British species.
48. „ ruficollis	
49. Cygnus buccinator	
50. „ americanus	

There is little then to mark Heligoland from the rest of Europe, so far as its list of rare migrants is concerned—all that we can say is, that it has been a little better worked and much more closely watched. The lost birds that wander to the British Islands from time to time are not one iota less interesting, or their occurrence less wonderful, either in the distance they have travelled or the route they have followed. That Heligoland is situated on or near one of the most important migration routes in the Eastern Hemisphere cannot be questioned; but there is nothing abnormal about migration there; and in some species the Flight is much more apparent in our islands. The most remarkable feature is, that migration becomes more contracted there than on the British or Continental coast-line, probably owing to Heligoland being so isolated, some twenty miles from land, offering not only an easily recognized land-mark, but a resting-place for lost and weary strangers from afar.

In one way Heligoland has the advantage of the British Islands. It is more favourably situated for tapping that great East to West migration wave (notorious for the comparative abundance of small Passerine species abnormally carried with it) that sets in from the far East in autumn and gradually spends itself down the Baltic and the valleys of the Danube, the Elbe, and the Rhine. Consequently many of the lost and wandering birds in the wrong stream of migration, are caught before they can reach our islands, or are more apt to be overlooked in a country so large as Britain. Moreover nearly all these Heligoland rarities are small Passerines, and of dull colours, or closely resemble commoner species whose presence excites no interest. Some, however, do escape, probably by way of the Rhine, and reach our shores, among the most remarkable being the Needle-tailed Swift. We are also richer than Heligoland in lost wanderers from the south and from America, owing to the British Islands being the first land to be sighted on this side of the Atlantic, at least *half* the species in the above table being from that Continent. Such facts are only too obvious, and make the wonders of Heligoland, great as they are, appear far more legendary than real. It may be worthy of remark to call attention to the curious fact of so many rare and abnormal migrants occurring in certain districts. Many of our rarest visitants have been taken one after the other, both individuals as well as species, along certain lines or in certain spots, which seems to indicate that there is at least some method in their movements.

I trust I have now succeeded in my endeavour to show at least a few of the principal perils that surround the migrant birds. Mortality is high amongst them; and yet this heavy death-roll serves a wise and important purpose, as a Check upon the undue increase of birds so specially favoured as they who live almost in a perpetual summer, or under conditions that entail the very lowest minimum of privation.

CHAPTER IX.

THE DESTINATIONS OF THE MIGRANTS.

EVERY regular migrant has a definite Winter Home.
At one end of its fly-line the usual breeding-
grounds are situated; at the other end the locality
where it spends the winter, or, as in the case of
a very great number of instances, the place where
it enjoys all the luxury of a second summer. To
make no allusion to these northern and southern
destinations in a work devoted exclusively to the
Migration of Birds would be an unpardonable
omission.

Broadly speaking, in the Northern Hemisphere, a bird's winter quarters are more or less directly south of its breeding-grounds; whilst in the Southern Hemisphere they are situated to the north of them. Of course there are exceptions to this, as we shall learn in the two following chapters, when we come to deal with the two great seasonal movements; whilst the distance between the winter and summer limits vary to a great extent, according to the length of migration flight and its general direction. The two great summer quarters of migratory birds are situated in the Palæarctic and Nearctic regions. The former region embraces Europe (including Iceland), Africa north of the Great Desert, Asia Minor, Northern Persia, and the remainder of Asia north of India, and the Yangtse valley to Japan and the Aleutian Islands. The latter region comprises Greenland and the entire continent of North America to about north latitude 20° in Southern Mexico. The latter region is by far the poorest in what we may designate Temperate species, probably in the ratio of about two to one. This is owing to the very obvious reason that when birds were banished from the circumpolar zone, glaciation was not only far more severe in the Western than in the Eastern Hemisphere, but the land surface, and consequent accommodation for bird life, directly south of the eastern half of that circumpolar zone, was more than double the area of that directly south of the western half of that zone. As a natural consequence, double the number of species retreated south into Europe and Asia, than penetrated into North America; or if the numbers were about equal, competition between species being keener, owing to more restricted area, and climate being so much more unfavourable to avian life, a much higher rate of mortality resulted in a present avifauna of such comparative poorness. Amongst its most important gaps may be mentioned the entire absence of the thoroughly Palæarctic Sylvinæ. In every part of these two mighty areas, from the highest Polar land yet visited by man, down to Mexico, Algeria, Northern Persia, the Yangtse, and Japan, migratory birds come from the south in spring to breed, and return to the south in autumn to winter. The migration flight of course varies to a very great extent, reaching for 10,000 miles or more in the most northern species, and dwindling down to perhaps a couple of thousand, or even one thousand, amongst the most extreme southern species. It now becomes necessary, owing to different climatal conditions, to speak of each of these great regions separately.

The great winter quarters of Palæarctic migrants are in Africa, India, South China, the Siamese Peninsula, the Malay Archipelago, Australia, and New Zealand. Birds visiting the West Palæarctic region, at least as far east as the Urals and the Obb valley, normally winter in Africa and Turkey in Asia, but a few retire to India, and a few to

China and Malaysia, birds that have obviously increased their summer area westwards within comparatively recent time. Birds visiting the East Palæarctic region, from the Urals or the Obb valley to Kamtschatka and the Aleutians, normally winter in Arabia, Persia, and India, eastwards to China, and southwards to Australia and New Zealand; but a few in a precisely similar way retire to Asia Minor and Africa, birds that have extended their summer area eastwards within equally recent time. The winter quarters of West Palæarctic migrants may be divided into three well-marked regions or zones. The first of these zones is the Great Aquatic Zone, consisting of the North Sea, the Baltic, the West Atlantic down to Madeira, the Mediterranean, the Black and the Caspian Seas, the Red Sea, and part of the valley of the Nile. The second of these zones may be said to include the extreme southern portions of Europe, Turkey in Asia, Persia, the Arabian Peninsula, and all Africa north of the Equator. The third of these zones consists of Africa south of the Equator, including Madagascar. The first or aquatic zone is the great winter quarters of West Palæarctic migratory water birds, especially Swans, Geese, and Ducks (ANATIDÆ), Grebes (PODICIPEDIDÆ), and various Waders (CHARADRIIDÆ). As might naturally be expected, the land birds wintering in the extreme northern portions of the second zone, between South Europe say and the Atlas, are few. Among them may be mentioned the Ring Ouzel (*Merula torquata*) and the Kestrel (*Falco tinnunculus*); but those that winter from South Europe to the Soudan and the Equator number a great many species, such as the Whinchat (*Pratincola rubetra*), the Grass-hopper Warbler (*Locustella locustella*), the Willow Wren (*Phylloscopus trochilus*), and the Tree Pipit (*Anthus arboreus*); whilst those that winter exclusively south of Europe are even more considerable. Among them may be included the Redstart (*Ruticilla phœnicurus*), the Pied Flycatcher (*Muscicapa atricapilla*), and the Marsh Warbler (*Acrocephalus palustris*). The birds journeying exclusively to the southern zone are comparatively few, and include such species as the Swallow (*Hirundo rustica*), the Red-backed Shrike (*Lanius collurio*), and the Cuckoo (*Cuculus canorus*). The birds whose winter range extends through all these zones are few, among them being the Quail (*Coturnix communis*), the Common Sandpiper (*Totanus hypoleucus*), the Kentish Plover (*Ægialophilus cantianus*), and the Land Rail (*Crex pratensis*). Some of these birds even winter as far north as the British Islands, and as far south as the Cape of Good Hope! Whether this apparently anomalous fact has any connection with the Neutral Zone of migrants alluded to in a previous chapter remains to be discovered.

In the East Palæarctic region much the same state of things will be found; but owing to the influence of the Gulf Stream, the winter zone of aquatic birds extends much further north in the

West than in the East. Thus, for instance, while
we find many Ducks and Geese wintering off the
coasts of Scandinavia, and even in the Baltic on the
Atlantic sea-board of Europe, but few, if any,
frequent Lake Baikal, the seas round Kamtschatka
or the Sea of Ochotsk, situated in precisely similar
latitudes in Central Asia and on the Pacific sea-
board of that continent. The aquatic zone of East
Palæarctic birds may be said to be the lakes and
great rivers of India and China, the northern portions
of the Indian Ocean, and the China, Yellow, and
Japan Seas. For the same great climatal reasons—
the East Palæarctic region being so much colder
than the West—the winter zones of land birds not
only do not commence so far north, but extend
many degrees further south than South Africa, into
Australia and New Zealand. The northern winter
zone may be said to include Turkestan, Afghanistan,
India, Burma, the Siamese Peninsula, South China,
and the south island of Japan. A comparison of
the winter range of some extreme eastern Palæarctic species with that of others in the extreme
west, affords interesting evidence of the great influence exerted by the Gulf Stream on the migration
of birds. Take, for example, the Redwing (*Turdus
iliacus*), breeding in Scandinavia, say in lat. 65°, and
wintering in the British Islands only ten degrees
further south, and compare it with the Dusky
Ouzel (*Merula fuscata*), breeding in a similar
latitude in Eastern Siberia (lat. 65°), but wintering
in China and Japan, more than 1000 miles further

south than the Redwing has to travel to find suitable winter quarters. The absence of Gulf Stream
influence lowers the breeding-range of great numbers
of species in the far east, just as in the far west its
presence extends them. Numbers of instances might
be given of Palæarctic birds reaching their highest
or lowest breeding-range in Scandinavia, where that
warm ocean current increases the temperature to a
very remarkable degree. The Brambling (*Fringilla
montifringilla*) breeds only as low as lat. 60° in Scandinavia, but can find temperature suitable for this
function at least ten degrees lower in Eastern Siberia.
In the west, owing to Gulf Stream influence, it is
able to winter throughout the British Islands, and
even in South Sweden ; but in the far east it is
compelled to journey into China and Japan, and
thus increase its fly-line by at least 1000 miles !
The Sand Martin (*Cotyle riparia*) is a circumpolar
bird, and breeds nearly up to the North Cape
(about lat. 70°) in Scandinavia, but in Kamtschatka
not any higher than lat. 55° ; and in the Nearctic
region not beyond lat. 68°. In this northern winter
zone by far the greater number of East Palæarctic
land birds visit India and South China, the land lying
immediately north of the Equator being by far the
most patronized, just as we find to be the case with
West Palæarctic birds. The southern winter zone of
East Palæarctic land birds includes the Malay Archipelago, Australia, and New Zealand, and curiously
enough is visited in the extreme south by comparatively few yet very similar species as in the west.

o

We now proceed to discuss the Nearctic region. Owing to prevalence of a much more severe climate throughout the northern portion of this vast land area, the comparative percentage of migratory birds is considerably higher than in the Palæarctic region. Although the land surface of the Nearctic region is, broadly speaking, only about half that of the Palæarctic region, the surface exposed to an Arctic climate (say with a midsummer mean temperature of 60° or less) is certainly more than double the surface exposed to a similar temperature in the Palæarctic region. This important fact is reflected in the migrations of Nearctic birds. The great majority of species breeding north of the United States, or in about the same latitude as the south of France and the valley of the Danube, are migratory; only the most boreal species remain to brave the rigours of a British North American winter. On the other hand, the summers are hot, and the entire region supplies breeding-grounds to a very large number of species.

The first or winter zone of aquatic migratory birds in the Nearctic region may be said to include the Great Lakes and rivers of the United States, and the St. Lawrence, down to the Gulf of Mexico in the Atlantic, and the Californian Gulf in the Pacific. The second or winter zone of land birds may be said to include all but the most northerly of the United States, Mexico, Central America, the West Indies, and the northern portions of South America, say to the Equator or the valley of the Amazon.

The third winter zone consists of the remaining portion of South America. It will be remembered that here, as in the East Palæarctic region, the winter zone extends many more degrees south than in the West Palæarctic region, due mainly to the severe or mild winter climate in the north of each area respectively. By far the greatest number of terrestrial species winter in the second of these zones, as we should naturally expect to be the case, seeing that the *Temperate* area of the Nearctic region is so remarkably contracted. Many species, however, especially among wading birds, penetrate far below the Equator, and some few extend their fly-line to Patagonia. Unfortunately, the meagreness of our knowledge of Neotropical ornithology is a serious obstacle to the exact definition of this great Southern Zone of Winter Migrants from the Nearctic region.

All these southern zones are eminently suitable for the winter residence of migratory birds. In every zone the various species wintering therein find food in abundance. Nowhere does any of these zones encroach upon a region of perpetual winter snow and frost, whilst the more tropical of them enjoy a climate beautiful in the extreme. The habits of the Migrants have been little studied in these winter quarters; but no bird breeds a second time within them during absence from its northern home. Many singing birds appear to regain their song, just as the Robin and the Starling do in our islands during winter; but Love and all that pertains thereto seems to remain utterly in abey-

ance. The winter is passed in a state of rest after
the mercurial energy of the previous summer, and
the fatigue of the long Migration Flight.   During
winter in these southern regions the land is a land
flowing with milk and honey for migratory birds;
insects swarm, the strands and mud-flats teem with
living things; the air, the water, and the vegetation
alike abound with food; whilst in more temperate
regions the ground is free from frost, and furnishes
birds that seek their food therein a seldom failing
store, and the berries and seeds on which many
birds almost exclusively fare are never hidden by
the snow for long together.   The great attraction
in these winter haunts is this superabundance of
food; whether it is sought in South America, in
Mexico, or the West Indies; in sultry Africa, or
India; in China, in Japan, in Malaysia, or far-away
Australia—it has been and is one of the strongest
incentives to the Migrants' southern flight during
the long-past ages that Avian Season Flight has been
the dominating necessity of its life!   But with the
return of the northern summer, as the sun steadily
progresses towards Cancer, a great change comes
over the scene.   Hot desolation fills many of these
southern lands; rivers and streams dry up; vegetation is scorched, and conditions of life for the
love-sick migrant hosts become as unbearable as
they became in the northern world in autumn, and
the great northern exodus begins.

So far the destinations of migrant birds have
been exclusively confined to those of species that

retreat from a northern winter and advance more or less closely towards regions enjoying a southern summer or a perpetual warm climate. It remains for us now to glance briefly at that other class of birds, not certainly so numerous, yet just as important, that is retreating from a southern winter and seeking a destination more or less closely towards a region where northern summer is prevailing, or climatal conditions are less rigorous. We have already had occasion to remark upon the comparatively small number of Southern Hemisphere Migrants, and to trace the cause; still the movement north is not only a marked but an excessively interesting one, and the Destinations of these northern winter migrants therefore demands notice. A migration from south to north only takes place upon the three great southern land masses of South America, Africa, and Australia, and from various small islands in the Southern Seas. With the sole exception of a few birds breeding on what we may call the outskirts of the glaciated Antarctic continent, this migration, for reasons we have already dwelt upon, does not extend far, if at all, beyond the Equator.

By far the most interesting example of this northern winter flight is that taking place in the Australian region, for there we have, so far as land birds are concerned, the only known important instances of birds absolutely crossing the sea to reach northern winter quarters situated in New Guinea and other islands of the Malay Archipelago.

In South Africa the northern zone of winter migration extends through the Transvaal and Damara Land to the Congo and the great system of equatorial lakes at the source of the Nile. In South America the northern destination of the Neotropical migrants reaches through La Plata to Peru and Brazil; and here, but on a much smaller scale, we have a sea flight north from the Falklands and Tierra del Fuego. Precisely the same causes influence migration in the Southern Hemisphere as in the Northern Hemisphere, scarcity of food owing to a prevailing lower temperature; although nowhere is this as acute as in the Northern Hemisphere.

With the return of spring in either Hemisphere the great wave of bird-life once more begins to flow north or south, according to circumstances, to the breeding-grounds in more temperate climes—a vast tide of Migrants that permeates every district and only spends itself in regions as far north or south as land is known! The great variation in climate over so vast an area of course determines the migration limits of individual species, due regard being paid to the isothermal lines of temperature. I am of opinion that much light will yet be thrown on the Phenomenon of Migration generally when this important condition is more closely studied. That Temperature has a great effect on a choice of breeding-grounds is unquestionable: that Temperature varies greatly over wide areas in obedience to ocean currents and other influences irrespective of latitude or longitude is equally certain. I only allude to this subject again in the hope of directing research thereto; it involves a much greater amount of labour than I have yet unfortunately been able to devote to it; and would, to be dealt with at all adequately, make far too wide a demand on the space here available for its discussion. After all, this little volume is but a pioneer in an almost unknown land, and no one is more conscious than its author of the utter impossibility of exhausting in a first attempt so wide and so little worked a subject as the Migration of Birds, even though that first attempt embodies the research and observation of a life-time. Its aim and purpose is but to point the way to more elaborate and detailed investigation, and to seek to rescue from chaos a branch of Ornithological science as fascinating and as absorbing as any alchemist's endeavour to accomplish the transmutation of gold !

That variation in Temperature was one of the most powerful initiating causes of Migration we have already seen; it is only natural to presume that it still continues to exert incalculable influence on Migration; otherwise we should not witness all these intricate and complex phenomena that so large a part of this volume has been written in the earnest endeavour to describe and possibly to explain. Depend upon it, the ebbing and the flowing of this great tide of Avian Life is not governed by chance; the habit of Passage is too vitally important, too deeply rooted, too grandly ancient, to have a trivial cause, either to be lightly acquired or to be readily relinquished.[1]   The present chapter, then, meagre as it is, will, I hope, serve to show that the Destinations of the Migrants are of some importance in a study of Avian Season Flight, and assist in no small degree to its ultimate elucidation.   Our knowledge of the geographical distribution of birds is still far from being even approximately complete.   Too little attention is paid to the habits of birds, or to the fact of birds being migratory or not, and, if the former, the dates of their arrival and departure, and the various causes that initiate these periodical movements.   Alas! only too often has an interesting and pregnant thread of investigation been suddenly snapped by the failure to obtain such requisite information ; and until we are in possession of this data I can confidently assert that much will remain unsolved and inexplicable in the Migration of Birds.

---

[1] The notoriously early age at which the Impulse to migrate is manifested, seems to me one very convincing proof of the vast antiquity of the habit of Migration.   Young birds, as we have already seen, are almost universally the first to display a restless desire for Flight in autumn, and to leave their birthplace before the bulk of the older individuals.

CHAPTER X.

THE Great Spring Migration of Birds may be said
to commence when the sun has performed about
a third of his journey towards the Tropic of Cancer,
or, in other words, about the middle of February.
Migration flows and ebbs with the sun. The
spring migration advances in the wake of the sun,
on his apparent northward course, and in the same
way retreats to follow that great central luminary to

the Southern Hemisphere. Unquestionably the one grand dominating impulse of Migration in spring is Reproduction. Migratory birds come north to breed, to rear their young in a climate where the temperature is best suited to their several requirements. That this is so seems proved by the fact that the adult birds are the first to migrate northwards in spring, birds whose sexual instincts are mature and strong; the young of the previous year in a great many cases do not extend their spring flights quite to the usual breeding-grounds of their species, and in other cases actually remain close to their winter quarters right through the summer. Another very remarkable fact about Spring Migration is the much greater rapidity with which it is performed. Birds may not fly any faster in spring, but they do not linger so long on the road; they seem bent on getting to their summer quarters as quickly as possible when once they have fairly started. It seems a wonderful and very remarkable fact how the earliest spring migrants time the date of their arrival to such a nicety that they are back in their old haunts almost to the moment that winter finally departs. But this fact is really not so very amazing after all. Birds as they near their northern destinations have often to wait about for spring, having been too eager to press onwards, and then great numbers collect on the very outskirts of retreating winter, ready to renew their flight at the first suitable opportunity. This is not so readily remarked in

such temperate latitudes as the British Islands
(although instances are on record where the earliest
flights of Migrants have been delayed for days
together in countries immediately to the south of
them), where the transition from winter to spring
is very gradual. In the Arctic regions, however,
season change is much more rapid, winter merging
into spring after a south wind of twenty-four hours'
duration; and then the interesting sight can be
repeatedly witnessed of birds arriving a day or so
too soon with the first signs of a thaw, and having
to retire south again for a few miles to the nearest
open water, or land free from snow. It is also
interesting to remark that many of the birds that
have the furthest to go are the last to start from
their winter quarters, and this appears to apply
equally to individuals as well as to species. Thus
the Swallows (*Hirundo rustica*) that breed in South
Europe begin to leave South Africa about the
middle of February, but those that breed in North
Russia delay their departure until the middle of
April, just as if they were perfectly well aware that
their summer quarters in the Arctic regions would
not be ready for them for several months longer!
This great wave of spring migration lasts practically
for about four months; beginning to set in towards
the north from the Antipodes about the middle of
February, and continuing until nearly the middle
of June, spending itself gradually at the latter
date in the highest Polar regions visited by birds.
Gradually this great migration wave may be traced

spreading northwards, from Africa, India, or China, for instance, right across Europe and Asia to the Arctic regions. All the various important Routes of Migration are more or less thronged with journeying birds; and at the intermediate stations famous for Flight, species after species appears at its usual time and passes on. At Gibraltar, along the chain of the Atlas mountains, at Malta, in the Greek Archipelago, in the Nile Valley, along the Red Sea, and at the great passes of the Himalayas, as well as up the Chinese river valleys and coast-lines, Birds in countless hosts are quietly pressing on— all with one common purpose in view, that of reaching their nesting-places as quickly as possible and settling down to family duties. As each species, or the individuals of each species, reach their haunts, the Great Avian Wave gradually decreases; the further and further north it flows the number of species become less, until all but the most Arctic ones remain.

Some species for reasons at present quite inexplicable follow a different route in spring from that which they traverse in autumn. This is proved by the following facts. The Nightingale (*Erithacus luscinia*) passes over Heligoland in April and May, but has never been caught there in autumn; the Dotterel (*Eudromias morinellus*) is rarely or never seen in Malta in spring, but passes that island regularly enough in autumn; the Turtle Dove (*Turtur auritus*) passes Heligoland commonly in May and June, but is much less abundant in autumn. The Whimbrel (*Numenius phæopus*) is another instance. In spring it passes the British Islands on passage much more abundantly than in autumn, and it has been remarked to fly much higher at the latter season. The intensity of the migration of each species varies considerably. First a few stragglers appear, and then the individuals gradually become more and more numerous until the migration is at its height; then it slowly dies down again, and a few laggards bring up the rear and conclude the Passage of each particular species for the season. At Gibraltar, for instance, the Turtle Dove begins to pass north on passage about the middle of April, and the flights gradually increase in volume until the first week in May, during which period the migration of this species is at its height, and then the numbers again decrease, until by the middle of the month the spring passage is practically over. At the same locality the Common Sandpiper (*Totanus hypoleucus*) begins to arrive early in March, and continues to increase in numbers until the middle of April, when it literally swarms there, after which the passage gradually gets less and less, until early in May, when it has ceased for the season. At Heligoland, at Malta, in Greece, in Asia Minor, at every station where migration has been carefully watched, precisely the same conditions prevail, and almost every species is alike in this respect. It is also a curious and interesting fact that some species migrate by night in spring,

yet travel by day in autumn, as for instance the
Quail (*Coturnix communis*); whilst others journey
by day in spring and by night in autumn, as for
instance the Common Bee-eater (*Merops apiaster*);
others yet again always by day or by night, or by
day and night at both seasons. The reason for
this change according to season remains a mystery,
and the practice may be found to prevail more
extensively when Migration becomes better known.

The general direction of Spring Migration, as
previously remarked, is broadly speaking from
south to north, yet there is a considerable amount
in other directions. Some of these what we may
call minor streams of migration are intensely interesting, not only in themselves, but as showing how
certain species have increased their areas of distribution in summer in a longitudinal rather than
in a latitudinal direction. Hence we not only find
migration from east to west in spring, but from
south-east to north-west, as well as the almost
universal stream from south to north. The Rose-
coloured Pastor (*Pastor roseus*) winters in India,
and migrates to South Siberia, Turkestan, the
Caucasus, South Russia, and as far west as Italy,
to breed. Canon Tristram observed incredible
numbers of this species on migration in spring,
crossing the plains of Syria in the Orontes Valley.
For three days during the last week in May flocks
continued to pass one after the other all flying due
west towards Europe. The birds made considerable noise as they flew along, chattering to each

other; and the sound of myriads of voices was
deafening as the clouds of birds passed on, or
wheeled and gyrated in the air like Starlings. A
vast flight of locusts had attracted the migrating
Pastors, and the birds formed "a great globe in
the air which suddenly turned, expanded, and like
a vast fan descended to the ground," which in a few
moments was covered with a moving black and
pink dappled mass of birds! On other occasions
trees were observed literally black with them; and
again, on being disturbed from a pool of water
where they had alighted to drink, the air was darkened with their numbers as they rose and hurried
out of harm's way. It is said that these vast flights
of Pastors are only seen in Palestine during spring
migration, when their flight is from east to west.
The Scarlet Rose Finch (*Carpodacus erythrinus*)
winters in India and Burma, yet extends its spring
migrations as far to the west as the Baltic Provinces,
and as far to the east as Kamtschatka. Its fly-
lines, therefore, go exactly north-east and north-west
from a common winter centre! The Black-headed
Bunting (*Emberiza melanocephala*) also winters
exclusively in India, and in spring migrates nearly
due west across Afghanistan and Scinde, to breed
in Persia, Palestine, the Caucasus, Asia Minor,
Greece, Turkey, and Italy. All these birds are
remarkably late migrants, not arriving at their
European breeding-grounds before May. They
all appear to have a dislike to cross the sea on
migration, and this explains their exceeding rarity

in the British Islands. They appear to follow the narrow strip of land, including Persia and Turkey in Asia, which is bounded by the Mediterranean Sea and the Persian Gulf on the south, and by the Caspian and Black Seas on the north; and their fly-lines must actually cross at right angles those of species passing from Africa to Northern Europe by way of the Black and Caspian Seas and Volga and Ural valleys. Then again we have migrants from the extreme south-east of Asia, *en route* to the far north-west of Europe. The Rustic Bunting (*Emberiza rustica*) winters in China, and extends its spring migrations as far west as Finland. The fly-line of individuals breeding in Europe apparently crosses Mongolia and follows the valley of the Yenesay and the Obb, thence across the Urals into those of the Petchora and the Dwina. The Little Buntings (*Emberiza pusilla*) that breed in North Russia, probably winter in India. The north-westerly fly-lines of these two birds are crossed by the north-easterly fly-lines of such birds as the Sedge Warbler (*Acrocephalus phragmitis*) and the Willow Wren (*Phylloscopus trochilus*), many individuals of both these species leaving winter quarters in Africa or Persia, and migrating at least as far to the north-east as the valley of the Yenesay, and higher than the Arctic Circle. The Roller (*Coracias garrula*) breeds in Cashmere and winters in Arabia and Africa, not visiting India at that season!

South of the Arctic Circle the arrival of spring

birds is gradual, weeks and even months separating
the date of appearance of certain species. This
seems entirely due to the fact that spring in temperate regions lasts for about a couple of months ;
but in the Arctic regions, where winter passes into
summer with scarcely a day's spring between them,
the arrival of migratory birds is sudden and much
more simultaneous. Migration in these high
latitudes depends almost absolutely upon the break-
up of the ice, which may be a few days earlier or
later according to local influences. Two of the
most complete and graphic records of the spring
migration of birds in the Arctic regions are those
made by Mr. Seebohm ; the first in the valley of
the Petchora (in company with Mr. J. A. Harvie-
Brown) during the spring of 1875 ; and the second
in the valley of the Yenesay during the spring of
1877. Both Migrations are very much alike in their
general aspects and conditions. In each case little
migration was observed until the ice on the rivers
began to show signs of dissolution. Such Nomadic
Migrants as Bullfinches, Snow Buntings, and Redpoles were the first to make their appearance, birds
that had wintered close to the fringe of perpetual
winter snow. Soon after these birds arrived, the
Hen Harrier (*Circus cyaneus*) and the Merlin
(*Falco æsalon*) appeared upon the scene, evidently
having followed their prey from the south ; and
as soon as these little Finches moved further
north or into the forests their enemies followed
them. About a week later (May 10th in the

Petchora valley), and as soon as the snow had melted here and there from the river banks, and the first streams of open water were visible, the Shore Lark (*Otocoris alpestris*) and the Bean Goose (*Anser segetum*) arrived, followed the next day by the Hooper Swan (*Cygnus musicus*), Bewick's Swan (*Cygnus bewicki*), and the Siberian Herring Gull (*Larus affinis*). The day following, May 12th, rain fell, and the first regular migrant Passeres, the White Wagtail (*Motacilla alba*), the Meadow Pipit (*Anthus pratensis*), and the Redstart (*Ruticilla phœnicurus*) "entered an appearance." On the 13th of May, Ducks of various species arrived, and in their wake the Peregrine Falcon (*Falco peregrinus*). On the 14th, the Reed Bunting (*Emberiza schœniclus*) was identified; and the day following a large flock of Common Gulls (*Larus canus*) appeared on the rapidly-breaking-up river. The week of the Arctic spring, during which the thaw of the day was more or less refrozen at night, was now over, and on the 16th of May summer set in with startling suddenness. The flooded river choked with melting snow overflowed its banks; the ice broke up with a great crash, and migration progressed more merrily than ever. On the 17th of May flocks of Fieldfares (*Turdus pilaris*), Redwings (*Turdus iliacus*), and Golden Plover (*Charadrius pluvialis*) were observed, and Geese, Swans, and Ducks in increasing numbers migrated down the river towards their more northern breeding-grounds. The migration of the

Meadow Pipit was still in progress, and the Red-throated Pipit (*Anthus cervinus*) was remarked for the first time. The Snow Buntings and Redpoles had now disappeared from the streets of Ust Zylma; flocks of White Wagtails had taken their place, and amongst them the Gray-headed Wagtail (*Motacilla viridis*) was observed. The next day brought the Lapland Bunting (*Emberiza lapponica*), the Whimbrel (*Numenius phæopus*), and the Teal (*Anas crecca*); and the last of the Snow Buntings passed on to the north. The Willow Wren (*Phylloscopus trochilus*) arrived on the 20th of May, and the next day a Crane (*Grus cinerea*) passed over, flying north at a great height. On the 22nd of May, the Siberian Chiff-chaff (*Phylloscopus tristis*), a distinguished stranger all the way from India, arrived; and with it came the Skylark (*Alauda arvensis*), the Tree Pipit (*Anthus arboreus*), and the Stonechat (*Pratincola rubicola*). On the 24th of May, the Brambling (*Fringilla montifringilla*) arrived. Two days later Oyster-catchers (*Hæmatopus ostralegus*), Ringed Plovers (*Ægialitis hiaticula*), Wood Sandpipers (*Totanus glareola*), and Temminck's Stint (*Tringa temmincki*) appeared, and a solitary Swallow (*Hirundo rustica*) was observed. On the last day of the month the Little Bunting (*Emberiza pusilla*), another traveller from the far south-east, appeared. The first few days of June brought the Cuckoo (*Cuculus canorus*), the Great Snipe (*Scolopax major*), the Terek Sandpiper (*Totanus terekia*), and the Black-throated Diver (*Colymbus arcticus*).

In the valley of the Yenesay, about a thousand miles further east, but in nearly the same latitude, very similar phenomena were witnessed. The Snow Buntings and the Redpoles heralded the migrant hosts, and proclaimed the advent of summer and the general break up of the great river. A Swan was seen, however, on the 5th of May; on the 10th, a few Geese; on the 16th (ten days earlier than in the valley of the Petchora), a Swallow arrived. During the remainder of the month, however, little migration was observed, but flocks of Geese and Swans were seen from time to time flying south! These were birds that had been too eager to get on their journey, and were compelled to return to open water far to the south. Summer was at least a fortnight later in Siberia than in Russia, and the Yenesay did not succeed in bursting his ice fetters until the first of June. With the break of the ice on that date migratory birds began to arrive in force, following in the wake of the gradually thawing river. "Although the first rush of migratory birds across the Arctic Circle was almost bewildering," writes Mr. Seebohm, " every piece of open water and every patch of bare ground swarming with them, a new species on an average arriving every two hours for several days, the period of migration lasted more than a month. Very little migration was observable until about the 22nd of May, although a few stragglers arrived earlier, but during the next fortnight the migration was pro- digious. In addition to enormous numbers of

Passerine birds, countless flocks of Geese, Swans, and Ducks arrived, together with a great many Gulls and Terns and Birds of Prey."

On the 1st of June, the first small insectivorous bird appeared, the White Wagtail, and the Brambling arrived almost simultaneously. From the 2nd to the 11th, flocks of Shore Larks continued to pass north; on the 3rd, the Wheatear arrived; the 4th brought the Dusky Ouzel (*Merula fuscata*), the Lapland Bunting, the Yellow-headed Wagtail (*Motacilla citreola*), and a great rush of Willow Wrens of no less than three different species, viz. the Siberian Chiff-chaff, the Common Willow Wren, and the Yellow-browed Willow Wren (*Phylloscopus superciliosus*). On the 5th, the Redwing, the Gray-headed Wagtail, and the Cuckoo arrived; whilst from the 1st of June onwards thousands of Swans passed over, all steadily flying north down the course of the great valley to the tundras beyond forest growth. The observations respecting the Bean Goose are particularly interesting, inasmuch that they illustrate very forcibly the progress of Migration in the Arctic regions. "Whenever," says Mr. Seebohm, "the weather was mild during May, small parties of Geese flew over the ship in a northerly direction. When the wind changed and brought us a couple of days' frost and snow, we used to see the poor Geese migrating southwards again. The great annual battle of the Yenesay lasted longer than usual the year that I was there. We had alternate thaws and frosts during the last three weeks

of May. Summer seemed to be always upon the
point of vanquishing winter, but only to be driven
back again with redoubled vigour. During all this
time there must have been thousands and tens of
thousands of Geese hovering on the skirts of winter,
continually impelled northwards by their instincts,
penetrating wherever a little open water or an oasis
of grass was visible in the boundless desert of ice
and snow, and continually driven southwards again
by hard frosts or fresh falls of snow. It was not
until the ice on the great river broke up that the
great body of Geese finally passed northwards"
(*Ibis*, 1879, p. 158). From the 5th to the 19th
of June many other small Passerine birds arrived.
On the 6th, the Red-throated Pipit and the Scarlet
Rose Finch (*Carpodacus erythrinus*) appeared; on
the 7th, the Dark Ouzel (*Merula obscura*) and the
Little Bunting; on the 8th, the Fieldfare and the
Lesser Whitethroat (*Sylvia curruca affinis*); on the
9th, the Sand Martin (*Cotyle riparia*) and the Yellow-
breasted Bunting (*Emberiza aureola*); on the 11th,
the Siberian House Martin (*Chelidon lagopoda*); on
the 13th, the Reed Bunting; on the 15th, the Sedge
Warbler; and on the 19th, the Mountain Accentor
(*Accentor montanellus*). During this period the
great migration of the Waders took place. "The
Common and Pin-tailed Snipes (*Scolopax gallinago*
and *stenura*) were," writes Mr. Seebohm, "the first
to arrive in company with the Asiatic Golden
Plover (*Charadrius fulvus*), on the 5th. The Wood
Sandpiper (*Totanus glareola*) and Temminck's Stint

(*Tringa temmincki*) arrived on the 6th. The
Golden Plover (*Charadrius pluvalis*) arrived on the
7th, and the Ringed Plover (*Charadrius hiaticula*)
with the Terek Sandpiper (*Totanus terekia*) on the
8th. The Ruff (*Totanus pugnax*) and the Dotterel
(*Charadrius morinellus*) arrived on the 9th; the
Great Snipe (*Scolopax major*) on the 11th; and the
Common Sandpiper (*Totanus hypoleucus*) on the
12th. On the 15th, the Green Sandpiper (*Totanus
ochropus*), the Red-necked Phalarope (*Phalaropus
hyperboreus*),and a solitary Curlew Sandpiper (*Tringa
subarquata*) arrived. Although migration continued
until the end of the month, during which many
new species of Passerine birds arrived, I did not
add a new species of Charadriinæ bird to my list
until we reached the tundras beyond the limit of
forest growth."

It would appear, however, that at least one-third of
these migrants visiting the North Russian tundras
in summer, vary their route somewhat, and do
not pass through Ust Zylma, coming by way of
the Baltic and White Seas, and thence along the
coast to the Petchora delta, leaving the great river at
its junction with the Ussa, instead of making the
long *détour* south-west which the course of the
Petchora here follows, and pushing along the valley
of the tributary to the tundras further north; perhaps even passing north direct, and leaving the river
system entirely. Seven species of Waders apparently
took one or the other of these alternative routes; two
Passeres and three Gulls (LARIDÆ) did likewise.

The possible explanation is, that the individuals of most of these species breeding on the Russian tundras winter more to the south-west, when their normal route would be through Western Europe; or travel from the far south-east by way of the Yenesay and Obb Valleys and across the Urals.

The same sudden character of Arctic migration in spring has been remarked by observers in much higher latitudes. The species are fewer, but they invariably make their appearance simultaneously with the final triumph of summer, an event which takes place even a little later in the season. Captain Feilden, when wintering in the *Alert* in Grinnell Land (lat. $82\frac{1}{2}°$), first observed the Knot (*Tringa canutus*), the Sanderling (*Calidris arenaria*), and the Turnstone (*Strepsilas interpres*) on the 5th of June. Knots were observed, however, a few degrees further south, in the winter quarters of the *Discovery*, on the 31st of May, which practically means that the birds here as elsewhere follow in the wake of retreating winter. On the 9th of June (in lat. $82\frac{1}{2}°$), Feilden observed the Brent Goose (*Anser brenta*) for the first time. In the Arctic regions of North-west America similar reports on spring migration have been made. Thus at Fort Simpson, situated at the junction of the Liard and Mackenzie rivers, about 150 miles north-west of the Great Slave Lake, Mr. R. G. McConnell reports on the spring arrivals of 1888: "The warm weather which commenced on the 1st of May continued throughout the month, and under its influence the snow quickly disappeared, and the spring advanced with astonishing rapidity. On the 20th of April, the first day the temperature rose above freezing-point for nearly six months, the Barking Crow (*Corvus americanus*) made its appearance. The Raven (*Corvus corax*) had remained throughout the winter. On the 1st of May, some Canada Geese (*Branta canadensis*) were seen at the edge of an open place in the river accompanied by a flock of Mergansers and other Ducks. The 4th brought the Robin (*Turdus migratorius*), and some Sparrows, and on the 5th the Wavies (*Anser hyperboreus*), which usually lag a few days in the rear of the Canada Geese, commenced to wing their way northwards, and in a couple of days were passing in such numbers that flocks were rarely out of sight."

The following table will give some slight idea of the spring migration of birds, half a dozen familiar species having been selected for the purpose of showing the gradual northern movement. (See p. 218.)

The subject of spring migration can scarcely be said to be exhausted until the probable route, and the dates of at least half a dozen stations to denote progress and duration, of every Palæarctic and Nearctic species have been carefully recorded. In the almost complete absence of such data the present chapter can only be regarded as a fragment; but sufficient I hope has been shown to denote the general characteristics of the Spring Migration of Birds, and at least to excite the reader's curiosity, and possibly stimulate him to

SPECIES.	CENTRAL OR SOUTH AFRICA (DEPARTURES).	GIBRALTAR AND SOUTH EUROPE (ARRIVALS).	BRITISH ISLANDS (ARRIVALS).	SCANDINAVIA (ARRIVALS).	ARCTIC REGIONS (ARRIVALS).
1. House Martin (*Chelidon urbica*)	February to April.	February to April.	April.	May.	June.
2. Swift (*Cypselus apus*)	February to April.	March and April.	April and May.	Middle of May.	June.
3. Cuckoo (*Cuculus canorus*)	February.	March.	April.	May.	June.
4. Nightjar (*Caprimulgus europæus*).	Feb. and March.	April and May.	Middle of May.	June.	
5. Common Sandpiper (*Totanus hypoleucus*) ...	February to April.	March and April.	April.	May and June.	June.
6. Sanderling (*Calidris arenaria*)	Beginning of April.	April and May (Passage).	April and May (Passage).	April and May (Passage).	June.

record not only his own observations, but to urge others, suitably placed, to go and do likewise.

From a study of the various facts enumerated in the present chapter, we may gather that Spring Migration is undertaken from an irresistible impulse to breed in a suitable temperature, so far as concerns Species, and that Individuals, with an inherent love for their birthplace and their home, seek such conditions in certain localities each season, returning unerringly to their old haunts, by the old route, such being the only one they know how to follow. Thus the pair of Willow Wrens that nest in a sheltered nook in the distant valley of the Yenesay return just as surely to their old summer quarters so long as life exists within them, and they are able to fly the long journey from Africa; just as surely as the pair of birds that breed amongst the bilberries on the rocky side of a Yorkshire coppice return to their summer home, although the two localities are more than 3000 miles apart! Yorkshire is the Mecca of one pair—of many pairs—just as the Yenesay or any other special haunt within the area of the Willow Wrens' distribution becomes the Mecca of all the rest! Not only are the fly-lines to each great centre different, but the necessities of the journey are utterly dissimilar; and the Willow Wrens breeding in Siberia could no more find their way to English haunts, than could a Sun Bird from South Africa find its way normally to the Yorkshire hills. Migration in Spring may

be said slowly to follow the retreat of winter; in
southern latitudes it is much more gradual and
broken, just as the grand seasonal change is slower
and the summer is long; in higher regions it is
rapid and continuous, because seasonal change is
quick and the summer is short.   In southern
regions most migrants are in no great hurry to
breed after their arrival—in the extreme south
they are remarkably late (later in the basin of
the Mediterranean, for instance, than in England);
but in the Arctic zone the pent-up sexual passions
reach their highest pitch before the summer haunts
are open.   The young of the Common Sandpiper,
for instance, are hatched in England before other
individuals of this species reach their Arctic breeding-grounds; so that the moment migrants arrive
in the high north they generally begin to breed.
If they did not do so, the object of their visit
would result in failure.   Mention has also been
made of the route of Migration varying in spring
and autumn; of the change from diurnal to
nocturnal flight according to season; but the
data on which these statements rest is remarkably
meagre, and suggest a wide and useful field for
future investigation.   In fact the work before the
student of Migration is endless, and as enthralling
as it is eternal.   Our space utterly forbids us to
do more than allude to the southern flight of
birds during the Antipodean spring, a subject
nevertheless of intense interest and importance.

CHAPTER XI.

THE AUTUMN MIGRATION OF BIRDS.

Impulse of Migration in Autumn—Conditions of Migration in Autumn reversed—Migrations of Cuckoos—Duration of Autumn Migration—Difficulty of Noting the Departure of Birds—Direction of Migration Flight—Gregariousness of Birds in Autumn—Autumn in the Arctic Regions—Migration Correlated with Terrestrial Change—Abundance of Birds in Autumn—Minor Streams of Autumn Migration—The Great East to West Wave of Western Palæarctic Migration—Rare Visitants to Western Europe from the East—Concurrent Streams of Migration flowing in opposite directions along the same Routes—West to East Migration in Nearctic Region—Punctuality of Departure in Autumn—Wandering Mania in Young Birds—Table of Autumn Migrants—Reversal of Migratory Impulses in Old and Young.

THE Great Autumn Migration of Birds is just as interesting, although it may not be quite so apparent in its earlier stages, as that taking place in spring. The most important impulse of Autumn Migration is unquestionably the failure of food supply; that this Impulse is even more deeply rooted than that which prompts migration in spring is apparent in the fact that All migratory birds, young and old alike, bow to its irresistible dictates. So often and so regularly has the grand southern Flight been performed, that the impulse is decidedly an hereditary one, as is proved by the fact that young birds and birds whose sexual instincts have been accidentally suppressed, are the first to obey its promptings and to pass south in autumn, just as we have seen that sexual passion being the strongest only in adults sends them first to the north in spring, and that in many cases (if not in all species where the young do not breed in their first spring) the young lag behind. The terrors of that far-off Ice Age, the dismay attending the banishment of Birds from the Polar world, have apparently been so deeply impressed upon migrants, that they have become hereditary terrors—an impulse, a restless longing desire, even in the young and inexperienced, to hurry away to warmer regions at the first possible moment. Curiously enough this impulse appears to be the most strongly developed in birds that breed the furthest north, where the extremes of an Arctic climate are the most pronounced. And what is equally remarkable is, that as soon as these migrants have reached more temperate latitudes they begin to loiter about, and to approach their winter quarters in a more leisurely way, as if conscious that they had left those regions behind where sudden winter might overtake them. We may thus very fairly attribute the Autumn Migration of birds to a fall of temperature at the approach of winter in their breeding-grounds, just as we have seen that the Spring Migration

of birds is due to a rise of temperature in their
winter quarters at the approach of summer. On
the one hand, the rise of temperature curtails food
supply, as well as introduces conditions that render
Reproduction, so far as we can learn, either impossible or undesirable; whilst, on the other hand,
the fall of temperature and its consequent ice
and snow cuts off the food supply more or less
completely and compels a southern movement.

In autumn many of the conditions of Migration are exactly reversed. The birds that have
the longest journeys before them are the first to
start, leaving the high latitudes where they were
born, or where they have spent the summer, at the
earliest moment. Young Knots (*Tringa canutus*)
and young Gray Plovers (*Charadrius helveticus*)
begin to pass Heligoland and the British Islands
early in August, some even with bits of down
sticking to their plumage. Both these birds winter
far in the Southern Hemisphere, reaching Australia,
South Africa, and South America, with fly-lines
more than 10,000 miles in length! Young Sander-
lings (*Calidris arenaria*) sometimes arrive on the
British coasts at the end of July; their fly-line is
equally lengthy, extending to South Africa, South
America, and the Malay Archipelago. The Bean
Goose (*Anser segetum*), on the other hand, with a
fly-line only extending as far south as the Mediterranean basin, China, and Japan, breeds as far north
as land is known, yet migrates later, and does not
reach its most southern quarters until winter is close

behind it! The Cuckoo (*Cuculus canorus*), with his fly-line reaching from the North Cape to South Africa, migrates early, passing Heligoland in July. In this species, it might here be remarked, the old birds migrate before the young, an anomaly due to the bird's parasitic instincts, which free it from all parental duties, and allow it to set off very early for the south! The Great Spotted Cuckoo (*Cuculus glandarius*) is an equally early migrant, a parasite leaving its summer quarters before its offspring. Indeed it seems doubtful whether these Cuckoos would migrate at all, if it were not for the influence of temperature on the nidification of their eggs, for they just come north to lay them, early in spring, when small insectivorous birds are nesting, and very soon afterwards draw south again. But the Cuckoos that breed in the normal manner are equally normal in their migration. The Yellow-billed Cuckoo and the Black-billed Cuckoo (*Coccyzus americanus* and *erythrophthalmus*) of America are comparatively late autumn migrants, but they build nests and hatch their own eggs in the usual way. It is also interesting to remark that the summer area of Parasitic Migratory Cuckoos is dependent upon that of the various species which play the part of foster parents, and the Migration Flight is consequently affected by the same cause. Were no small insectivorous birds to visit the Arctic regions in summer, the Common Cuckoo would never have extended its range to the Arctic Circle, just as the narrower summer area of the

Great Spotted Cuckoo is controlled by the smaller geographical limits of the species selected to hatch its young and bring them to maturity. The Blackcap (*Sylvia atricapilla*), again, with its comparatively short fly-line, reaching from lat. 60° in Scandinavia down principally to North Africa, and more rarely towards the Equator, is a late migrant, passing Heligoland in October and November. The rule appears to be that the further north a bird breeds, the more anxious it is to get south in autumn, and the longer its fly-line the earlier it is to start. We have already alluded to the fact that some species reverse their routes almost entirely in autumn, and travel south by quite a different fly-line from that they followed north in spring; we have seen that the time in some species is reversed, and night instead of day is selected for Flight. Migration in autumn is also slow and leisurely in comparison with the wild mad rush of spring; birds take their time in moving south, dallying here and there for a few days wherever food is plentiful; taking it easy, as it were; enjoying a well-earned holiday after all the bustle and excitement of summer.

The Autumn Migration of birds in the Northern Hemisphere practically lasts about four months, beginning, we will say, during the latter half of July, and continuing until the first half of November. Some few species wander south later than this, but these must be regarded as Nomadic rather than Regular migrants. Autumn Migration is at its

Q

height during September and October, being much less pronounced towards the extreme dates or periods of its duration. From these facts, and those in the preceding chapter, it will be seen that Migration is more or less in progress during ten months out of the twelve, only lapsing for a period of two months in December and January. As we have already seen, however, even these two months are characterized by certain local movements, so that really migratory birds, generally speaking, can never be regarded as in a state of absolute rest. For various reasons the Autumn Migration of birds is not so observable as that in spring, at least so far as the actual time of departure is concerned. Of course when birds are on direct passage the movement is easily enough remarked along all the recognized routes, species after species passing certain stations with considerable punctuality. It is, however, a very difficult thing to note the exact period of departure from the place where the migration of a species begins. Most birds moult just previous to migrating in autumn, and at that time they become very skulking in their habits, lose their voice, and make a practice of slipping so quietly away that they are not missed perhaps for days after they have really taken their departure. The only reliable sign therefore that migration is in progress is that of individuals actually *en route* at a distance more or less remote from their starting-place. The great wave of Autumn migrants begins to flow south in July, gradually increasing in

volume month by month, until it is in the full
tide of its movement, and the temperate zones of
the Northern Hemisphere are once more teeming
with pilgrim birds. Now the general direction is
from north to south, but many pass from north-
east to south-west, and from north-west to south-
east, the fly-lines crossing each other just as intricately as in spring. In the Palæarctic region we
have also a very important stream of Migration
flowing nearly due West from the East; another and
a smaller stream setting in due East from the West.
We shall have occasion again to allude to these
various streams of migrants later on. The great
intensity of Migration Flight is much more sudden
in autumn than in spring, often great numbers of
birds appearing at once, not gradually becoming
more and more numerous; the ebb is about the
same. Another very remarkable fact connected
with Autumn Migration is, that the birds observed
on passage are much more abundant than in spring,
and that the tendency to fraternize among individuals, as well as among species, is greater.
Young birds especially are prone to this gregarious
habit, and of course it is of these young birds that
the great majority of autumn migrants consists.
As soon as ever the broods are sufficiently matured
to be independent of parental care, a gregarious
instinct becomes predominant, especially among
Waders, and flocks are soon formed at the common
feeding-ground which eventually start off on migration
in company. The old birds are delayed somewhat,

by having to complete their moult, but then they too, in a great many cases, pack together and migrate in company. Unfortunately we have not much reliable data respecting the autumn movements of birds in the Arctic regions just previous to migration, but in more temperate latitudes the information is ample. In the Arctic regions the great wave of Autumn Migration is not so apparent in its departure as in its arrival, but further south towards the grand winter quarters, when the flood-tide of this mighty host is gradually accumulating, it is even more apparent than in spring, the points of greatest influx being of course exactly reversed. The first signs of autumn appear in the Arctic regions during August, when the sun drops behind the horizon for a little time every twenty-four hours, his stay gradually getting longer and longer as September arrives. Then the frosts begin, and by the end of that month summer is banished for the year, and the great army of migrants that had collected here from almost every corner of the world, has fled south again, and the land is given up to silence and desolation once more. Species after species for two months has been speeding south; all the old winter quarters in temperate and tropic latitudes have been slowly filling with banished birds, fleeing from the terrors and the hardships of a northern winter; and with the gradual spending of this mighty wave of Avian Refugees the grand phenomenon of Bird Migration completes once more its annual cycle.

Another equally interesting fact concerning
Autumn Migration is, that all the more northern
birds that appear earliest at their breeding-grounds
in spring, as a rule, linger the longest at them in
autumn, probably because they are the hardiest and
most robust of migrants. The Wheatear (*Saxicola
œnanthe*) and the Chiffchaff (*Phylloscopus rufus*),
for instance, arrive amongst the earliest of birds
throughout their summer area of distribution,
reaching the British Islands during the last week in
March; they linger in the autumn until most
others have departed, being amongst the last to go.
On the other hand, the Swift (*Cypselus apus*) and
the Red-backed Shrike (*Lanius collurio*) arrive very
late in spring (in the British Islands not before
May), and are amongst the earliest to retire south
in autumn. Probably these very early autumn
migrants would be the first to relinquish a northern
journey in the event of the summers becoming
shorter than they are now, just as the hardy late
autumn migrants would continue to visit the
northern zones as long as any summer remained at
all; or, in the event of warmer climatal conditions
rendering the present winters shorter and milder,
they would be the first to forego migration altogether. This is another deeply interesting instance
of how some of the grandest changes our Planet
has undergone are indelibly stamped upon Avian
Migration, or are correlated with such a comparatively insignificant phenomenon as the Season Flight
of Birds.

The Autumn Migration of most birds is much more marked than in spring, and the vast flocks of certain species that may be regularly witnessed passing to their winter quarters exceed in numbers anything seen in the Vernal movement, even in the Arctic regions. For instance, the Little Bustard (*Otis tetrax*) is described as crossing the Caucasian Steppes on autumn migration literally in millions; the flights of Sky Larks (*Alauda arvensis*) at that season are almost past belief, crossing certain points for days and nights together in one scarcely broken stream. Prjevalsky observed the Needle-tailed Swift (*Chætura caudacuta*) on autumn passage in Mongolia passing overhead for an entire day almost without cessation. The vast hordes of Waders seen on passage are also most characteristic of autumn.

If the predominant direction of Autumn Migration in the Northern Hemisphere is from North to South, many minor streams take a different course. Unquestionably the most important of these is the one that sets in from the East and follows a course nearly due West, in the Palæarctic region. This peculiar, yet very marked, stream of Autumn Migration is as yet far from being perfectly understood, although its existence is a demonstrable fact. It is composed of many species of birds, such as Larks, Starlings, Thrushes, Crows, and Finches, that breed more or less abundantly in Eastern Europe and Western Asia, for the most part hardy species, not exactly obliged to winter in India, or even in North Africa, but forced to leave their

eastern haunts owing to the climate being much
more rigorous than in Western Europe, where the
influence of the Gulf Stream renders the winters
comparatively mild and genial. This great wave
of hardy autumn migrants probably begins to flow
westwards from the valley of the Yenesay, and
drains a strip of country, perhaps 1000 or 1500
miles wide, gradually spending itself in Western
Europe and the British Islands, where many of
these migrants pass the winter. The route followed
appears to be by way of the Aral and Caspian Seas,
along the lower valleys of the Volga and the Don,
the northern coasts of the Azov and Black Seas,
and the valleys of the Dnieper and Danube. Probably throughout its course contingents of birds
are from time to time leaving the main artery, and
flying south into Persia, Asia Minor, Turkey,
Greece, and Italy, as more or less important influxes
of the birds forming this stream are remarked at
that season in those countries. Many odd individuals of eastern species, whose proper line of
migration is south or south-east, get into this
western stream of birds and are borne into Western
Europe, some of them even reaching the British
Islands, Heligoland, and other localities. These
lost stragglers are very important indications of
this western Migration in autumn. Some of them
belong to southern species, but their summer area
extends for a considerable distance north in Asia,
say into Turkestan and the extreme south of
Siberia. Now no northern Migration is known in

autumn in the Northern Hemisphere, yet these
southern and eastern birds occur at that season,
say in Heligoland or the British Islands, far to the
north of any localities they may visit in summer in
South-western Europe. In addition to this evidence,
we have also the important fact that in cases where
a species is composed of two races, an eastern one
and a south-western one, the individuals that reach
us generally belong to the former race ; whilst on
Heligoland the total number of stragglers from the
south-east of Europe is far more than from the
south-west of that continent. There are, for
instance, two races of the Black-throated Chat, viz.
Saxicola stapazina, an inhabitant of the extreme
west of the Mediterranean basin, and *Saxicola
stapazina melanoleuca*, an inhabitant of the extreme
east of that basin, Asia Minor, and Persia. An
example of the eastern race has been shot in the
British Islands in October. Again, individuals
belonging to such thoroughly eastern species as
the Isabelline Chat (*Saxicola isabellina*), Richards'
Pipit (*Anthus richardi*), the White-winged Lark
(*Alauda sibirica*), the Black Lark (*Alauda tartarica*),
and Macqueen's Bustard (*Otis macqueeni*), are
occasionally brought on this western stream of
migration, not only to various parts of western
continental Europe, but to Heligoland and the
British Islands. None of these birds are migrants
to Africa ; their winter quarters are in South
Turkestan, India, Burma, and South China, but
stray individuals take the wrong direction in autumn

and join the western instead of the south-eastern
stream of birds. At each period of migration,
therefore, whether in spring or whether in autumn,
we have the curious fact presented to us of two
very important minor streams of migration con-
currently flowing from West to East and from East
to West along one common route ! Rose-coloured
Pastors, Rose Finches, and Black-headed Buntings,
for instance, on their way from Europe to India,
actually pass the western stream of migrants from
Asia, with odd White-winged Larks and Richards'
Pipits intermingled on their way to Europe ! These
bye or minor streams of autumn migrants are for
the most part a result of extension of area in the
species composing them, necessitating a more or
less pronounced longitudinal as well as latitudinal
flight. Doubtless many other such minor streams
remain to be discovered. There can be little doubt
that a similar stream sets in from Central Asia and
flows south-east towards the coast of North China,
the Corea, and Japan, composed principally of hardy
birds of eastern origin that have extended their
range westwards in summer into the colder climate
of Siberia. The Black-throated Divers (*Colymbus
arcticus*), for instance, that breed in Eastern Asia
migrate east in autumn along the Amoor valley
and other well-recognized routes of Passage, to
winter in the Japan Sea. The eastern form of
the Common Gull (*Larus canus niveus*) leaves
its Siberian haunts at the same season, and passes
along very similar fly-lines to the coasts of China

and Japan for the cold season. Many Carrion Crows (*Corvus corone*) breeding in Eastern Asia apparently also migrate due east or south-east in autumn to districts on the Pacific sea-board; and it is not improbable that individuals of this species even from South-east Russia and Turkestan take this eastern journey, seeing that the Emigrations of the Carrion Crow in past ages were apparently from a far eastern centre of dispersal, and that its nearest surviving allies are Oriental (*Corvus macrorhyncha*), and even Australian (*Corvus australis*). Probably vast numbers of Larks (the eastern races of *Alauda arvensis*: *Alauda dulcivox* and *cœlivox*) and other species migrate in autumn along similar routes, but the subject requires much further investigation before we can form any tolerably complete estimate. We also certainly find precisely the same phenomena in the Nearctic region, of birds having extended their area in a longitudinal direction, and migrating in autumn from West to East. Several species of American birds breeding in the Arctic regions of that continent cross Behring Strait into Asia in spring, returning to winter in the southern portions of the Nearctic or the Neotropical regions. Among these birds may be mentioned the Semipalmated Plover (*Ægialitis semipalmatus*), the western form of the Red-breasted Snipe (*Ereunetes griseus scolopaceus*), and the Buff-breasted Sandpiper (*Tryngites rufescens*). This East to West Migration will be further noticed in the following chapter.

Although the fact may not be so apparent as in spring, each species has its regular time of migration in autumn, to which it keeps with wonderful punctuality. Almost to the day birds may be missed in autumn just as we may look for them to arrive on a certain date in spring. This rule, however, can only be strictly applied to adults; the young birds migrate either as soon as they can fly, or begin to wander from their birthplace as soon as they can feed themselves. The movement is generally in the direction of the usual migration. Birds appear very regularly at certain points *en route* in autumn, but generally in more variable numbers than in spring. It is also worthy of remark that the young birds in certain northern species, especially in Nomadic Migrants, wander the furthest south. Everywhere young birds seem possessed more or less with a mania for wandering about. Some of the most extraordinary instances of birds occurring far from their usual habitat, have been those of young or immature individuals—as if on the look-out for new haunts in which to settle; and this, combined with the ordinary migratory impulse, often sends them far from their usual area.

As in the preceding chapter, the following table will give some idea of the Migration of birds in Autumn, the same species being selected to illustrate the return journey. As previously remarked, however, the winter quarters of the House Martin are very imperfectly known :—

SPECIES.	ARCTIC REGIONS (DEPARTURES).	SCANDINAVIA (DEPARTURES).	BRITISH ISLANDS (DEPARTURES).	GIBRALTAR AND SOUTH EUROPE (DEPARTURES).	CENTRAL OR SOUTH AFRICA (ARRIVALS).
1. House Martin (*Chelidon urbica*) ...	August and Sept.	September.	Sept. and Oct.	Sept. to Nov.	Oct. and Nov.
2. Swift (*Cypselus apus*) ...	August.	August.	August and Sept.	Sept. and Oct.	Oct. and Nov.
3. Cuckoo (*Cuculus canorus*) ...	July.	July.	July and August.	August and Sept.	Sept. and Oct.
4. Nightjar (*Caprimulgus europaeus*).		August and Sept.	September.	Sept. and Oct.	October.
5. Common Sandpiper (*Totanus hypoleucus*) ...	August and Sept.	August and Sept.	August to Oct.	August to Oct. (Passage).	Sept. to Nov.
6. Sanderling (*Calidris arenaria*) ...	August.	August and Sept. (Passage).	August to Oct. (Passage).	Sept. to Nov. (Passage).	Sept. to Nov.

From the above remarks we may conclude that the Autumn Migration of Birds is initiated by a fall of temperature and consequent failure of food, in conjunction with an unquestionably strong hereditary impulse to move south from regions where ages of accumulated experience have taught that a winter sojourn means death. Young birds seem more strongly impelled to migrate in autumn than in spring, seeing that in so many species they not only leave before their parents, but are more or less reluctant to quit their winter quarters the following spring. On the other hand, the impulse in old birds to migrate in spring is stronger than in the young, owing to sexual instincts being so much more highly developed. The impulse to migrate in young and old seems exactly reversed according to season. In Autumn the Young are the eager ones, the Old the laggards; whilst in Spring the Old are the eager ones, and the Young are the laggards. The Migration of Birds in Autumn follows in the wake of retiring summer and in the van of winter, just as in spring it follows in the wake of retiring winter and in the van of summer. Migration in Autumn in the Southern Hemisphere, so far as we can judge from the scanty details on record, is of a very similar nature to that prevailing in the Northern Hemisphere, and initiated by exactly the same set of causes, namely, failure of food in consequence of a fall of temperature.

CHAPTER XII.

MIGRATION IN THE BRITISH ISLANDS.

PERHAPS no other part of the world is more admirably situated for the study of Migration than the British Islands. Almost every known kind of

Avian Season Flight may be witnessed upon them.
They are the summer quarters of vast numbers of
birds; the winter rendezvous of others. They are
situated on the direct fly-lines of many northern
birds that pass over them in spring and in autumn
to breeding-grounds in the Arctic regions and to
winter haunts in the Tropics. They are visited by
birds that fly north during the Antipodean winter;
by birds that come from the far East, and not a
few from the trans-Atlantic West. Much Nomadic
Migration breaks upon them; whilst the rare birds
obtained on abnormal flight within their area equals,
if it does not exceed, in number and in interest, the
occurrences in any other part of the world. Not
only so, but Migration in the British Islands is of
an unusually diverse character, remarkably universal,
and subject to all those fluctuating incidents that
until the last few years were presumed to be peculiar
to one European station alone. The fact is, Migration was never studied in our islands at all, except in
the most cursory manner, until two British ornithologists, Messrs. J. A. Harvie-Brown and J. Cordeaux,
went to work in the only practical way, by enlisting
the aid of persons stationed at lighthouses and light-
vessels, and best situated for watching and reporting
the Season Flight of Birds. British naturalists are
very deeply indebted to these gentlemen, for it
is to their labours that the systematic study of
Migration in the British Islands is almost entirely
due. For several years unaided they published
returns on Migration from numerous stations round

our coast in the *Zoologist* and elsewhere. In 1881 the investigation was countenanced and authorized by the British Association for the Advancement of Science, and a Committee appointed to carry on the work, and make annual reports thereon. Gradually the area of observation has been extended, until the stations numbered upwards of two hundred. Nine Annual Reports have been compiled chiefly from schedules sent in by light-keepers, dealing not only with the customary spring and autumn migrations for every year, but with much evident Local Movement and Nomadic Migration. The result is an immense amount of information—of raw material, from which it may be possible to obtain some important facts, but the labour will require very great care and discrimination on the part of the compiler, who must also have a keen appreciation and extended knowledge of Migration as a whole. To attempt wide deductions on Migration Philosophy from this mass of raw localized material will end in failure. It is evident in many of these reports that too much importance has been attached to what are purely local and fortuitously initiated movements, and that many of the birds referred to are not flighting at all in any sense of the term.

Be all this as it may, however, we are now furnished with abundant evidence that Migration over the British Islands prevails to a very astonishing but previously utterly unsuspected extent. No longer does Heligoland stand alone in its Migration importance; scenes just as wonderful and just as frequent may be witnessed from scores of stations in the British Islands likewise. In many cases Migration is even more marked on our eastern coasts than on Heligoland. This is only just what we ought to expect. Had Heligoland retained its unique character as a Migration Station, much of the mystery in Avian Season Flight that is now made plain would only have been intensified. Many of these simple and obviously truthful British Migration Reports read like romance, and equal anything that has come from the famous bird rock at the mouth of the Elbe. Birds striking the lights; birds in countless hosts, drifting by in feathery tides; birds in hundreds exhausted and falling into the sea to perish, or allowing themselves to be taken by the hand; birds passing for days together, literally square miles of them; birds by day and birds by night, flying in regular steady waves or in bewildering rushes; birds following the rays of revolving lamps, or hurling themselves against the dazzling beacons to die, or settling in crowds to rest! Such are the facts observed by the light-keeper on his lonely vigil, and forming a chapter in the life history of our feathered friends of profound, intense, and marvellous interest! When we bear in mind that the records of Heligoland have been religiously kept by a distinguished naturalist for half a century, and that our reports are made almost in every case by untrained and unscientific observers, the possibilities of future research seem limitless.

To deal fully with the subject of migration in

R

the British Islands, it will be necessary to recapit-
ulate at least a portion of our earlier chapters. The
first portion of the subject on which it becomes
necessary to dwell is that of Routes. Undoubtedly
the most important Highways of Migration in the
British Islands are along the coast-lines. This is
very forcibly expressed by Mr. Cordeaux : "An
observer taking up his position at a short distance
from the coast would see or know nothing of
Migration, yet within half a mile or less there might
be a constant stream of birds, hour by hour, and day
by day, passing to the south." This is probably owing
to the small area of the islands, and the coast-lines
trending almost entirely from north to south—a
direction favouring passage not only for species
breeding or wintering in this locality, but for those
birds that pass along them during spring and
autumn flight. Undoubtedly the most important
of these extend along the east and south coasts,
and drain a considerable migration from Scandinavia
by way of the Shetlands and Orkneys. At the
latter, the route seems to branch into two, one
following the west coast of Scotland, including the
Hebrides, to the north of Ireland, where another
branch occurs, the western following the Atlantic
seaboard (a route little traversed by Passerine
birds) ; the eastern, the coasts of St. George's
Channel, the Irish Sea, and the Bristol Channel,
along which highway great numbers of small birds
pass from the south, and *vice versâ*. These Routes
may be roughly taken as the general direction of

migration, but they are subject to an immense
amount of local modification. Thus all or nearly
all the great indentations of the coast are skipped
by land birds (Waders keep much more closely to
the windings of the coast, especially such as afford
suitable feeding-grounds), which make a practice of
flying from one headland to another. The normal
fly-line, for instance, down the west coast of
England from Scotland misses all the English
coast-line of the Irish Sea, and crosses from the
Mull of Galloway, *via* the Isle of Man to Anglesey.
In the same manner many of the wide Scottish
Firths are crossed, as well as many of the more important promontories. Doubtless these practices are
general among migrants throughout the world, for
there is nothing to lead us to presume that the
British Islands are abnormal in this respect. One
or two local lines of Migration have also been
indicated, such as across the narrow country
between the Firths of Clyde and Forth ; and across
England from the Wash to the Bristol Channel,
onwards probably to Ireland.

Owing to the much more rocky character of the
east Scotch coast-line than that of the east of
England, the waves of Migration are more or less
compressed into the various Firths and river-valleys,
which serve as the passages to and from the inland
districts, such as the Pentland and Moray Firths
and the Firth of Forth. On the eastern coast of
England, where the sea-board is low, a more general
ingress and egress is made, although there is

certainly some evidence that the valleys leading
from the Humber, the Wash, and the Thames are
feeders for the more inland districts. That the
Humber and the Wash are favourite routes to the
interior, I have gathered much evidence to prove,
mostly from personal observation. The Goldcrests
(*Regulus cristatus*) that strike the Humber district
in autumn, sometimes in enormous rushes, may
be traced right up the valleys of the Don and
Sheaf and Trent almost to their source. The
Song Thrush (*Turdus musicus*) follows a similar
course, as also do various Waders and Crows.
The Hooded Crows (*Corvus cornix*) that in equally
large flights strike the Wash district may be traced
along all the river-valleys, the Witham, Welland,
Nene, and Ouse into the adjoining counties. For
some reason not at present quite clear, birds prefer
to enter or leave the country *via* a depressed coast,
often following below a long line of cliffs (perhaps
for shelter) until a more suitable district is reached.
This is essentially the case along the rock-bound
south coast of England, where Mr. Swaysland
assured me many years ago that migrants made
a practice of "cuddling the cliffs," and as I have
myself on various occasions witnessed both in
autumn and in spring. This was in the vicinity
of Brighton. Further west in Devonshire, I am
of opinion that the Dart valley is another important
gateway of migratory birds. From Dartmouth to
Berry Head the coast is very precipitous, and but
little migration appears along it ; but migrants

enter the Dart valley and spread over the surrounding country right up to the shores of Tor Bay,
often in surprising numbers.   I have traced the
migration flight of Cuckoos, Warblers, Redstarts,
and Flycatchers along this route in spring.   It has
also been remarked that migrants coming into our
islands in autumn seldom or never alight on the
coast after a favourable passage, but fly inland at
once, which I, myself, have often witnessed in the
Wash and elsewhere.   On the other hand, if the
journey has been rough and fatiguing, the migrants
are glad to drop on the nearest land, often tired
out and utterly exhausted.   I have seen Goldcrests
and Linnets (*Linota cannabina*) swarming in thousands in the Lincolnshire salt-marshes, many of
them so tired and exhausted as to fly with the
greatest reluctance.

Although I am very strongly of opinion that
all or nearly all our indigenous birds are more or
less migratory, what we will designate the British
migrants proper may be divided into three very
distinct classes.   First, we have the birds that come
here in spring to remain with us during the summer
and rear their offspring; second, the birds that
come here in autumn to spend the winter with us
and leave in spring; and third, the birds that only
pass our islands to and from the more northern
breeding - grounds   and   more   southern   winter
quarters.   Perhaps we might also add a fourth
class, consisting solely of Nomadic Migrants, whose
appearances are irregular and intermittent, yet often

very considerable, and in many cases composed of
individuals of various species driven from one part
to another by stress of weather.   The usual marvellous punctuality of arrival and departure, and persistency of following certain routes; the same order
of Migration; in fact all the minor details of
Flight are observed by British Migrants as we have
seen is so universally the case in earlier portions of
this work, so that they need not be more than
alluded to again.   We might also, however, remark
similar obedience to Meteorological changes and
choice of Flight wind.   The evidence so far as it
goes seems to imply that direction of wind is
subservient to a very great extent to change of
temperature, the latter, if adverse, initiating Migration even in the face of contrary or unfavourable
air-currents.   The consequence is, that the prevailing winds at the time of Flight, especially in
autumn, influence to a great extent the wave of
migrants, deflecting it in certain directions, and
causing it to be of a broad and expansive character,
rendering the tide gentle and continuous, or compressed and narrow, rendering it more throbbing
and in occasional rushes.

   We will now proceed to glance briefly at each
of these great classes of Migrants, beginning with
those in spring.   Migration at this season begins
almost as early in the British Islands as in South
Africa, but of course the species affected are very
different.   The first decided migratory movement
is noticeable, say in February, when various birds

that have been wintering in this area begin to pass towards the continent. The most noticeable of these are such species as Blackbirds (*Merula merula*), Thrushes (*Turdus musicus*), Redwings (*Turdus iliacus*), Fieldfares (*Turdus pilaris*), Pied Wagtails (*Motacilla yarrelli*), Meadow Pipits (*Anthus pratensis*), Larks (*Alauda arvensis*), Rooks (*Corvus frugilegus*), Hooded Crows (*Corvus cornix*), various Finches, such as House-sparrows (*Passer domesticus*), Linnets (*Linota cannabina*), Redpoles (*Linota rufescens* and *linaria*), and Snow Buntings (*Emberiza nivalis*). All these birds continue now at intervals to pass out of our islands for the next two months, sometimes leaving in considerable flights which last almost incessantly for days together; whilst as the spring advances, Starlings (*Sturnus vulgaris*), Goldcrests (*Regulus cristatus*), and other birds begin to move, and the great departure of Ducks (ANATIDÆ) and Waders (CHARADRIIDÆ) commences. Many of these birds compose the great autumn wave of Eastern Migrants, but the numbers that return in spring are rarely so marked. Instances of great vernal rushes, however, have been recorded. For example, it is reported from the Swin Middle Light-vessel, stationed some twelve miles off the Essex coast, that on the night of February 14th to 15th, 1885, great numbers of Larks were passing towards the south-east; ninety of them came on deck, numbers fell into the water, and " for two hours the Larks were like a shower of snow." These birds were evidently migrating out of England

towards the coast of Belgium, on the old route back
to the East. Another very important migration,
chiefly of these early migrants, is also, incredible as
it may appear, actually coming into England *from*
the Continent at the same period; so that we have
the astounding phenomenon of a marked and
constant Migration across the North Sea in exactly
opposite directions! These latter migrants are birds
that evidently breed in our islands, but for some
reason prefer to winter on the Continent. This
movement may be due to the influence of Temperature varying on individuals of the same species.
The Crows, Starlings, Larks, Buntings and Finches
spending their summer in Great Britain, may require
a higher winter temperature, and retire towards
Southern Europe to obtain it; whilst those from
the colder east and north area of continental land,
and naturally of more robust constitution, find the
temperature suited to their individual requirements
in our islands. In autumn this Cross Migration
may in part be due to the stream of migrants from
North-western Europe, which passes along our
coasts to the south, meeting the east to west stream
of migration, composed of birds on their way to
winter with us. Whatever the dominating cause,
the fact is unquestionable, and the Cross Migration
is not only regular but persistent, especially during
seasons remarkable for great and sudden changes
of temperature. I firmly believe that temperature,
as affecting individuals, is the chief initiating cause
of these singular Avian waves, and that their ebbing

and flowing is not voluntary, although Mr. Cordeaux appears to be of the opinion that it is. The necessities accruing in past ages, and handed down as hereditary impulse, may also be a minor cause.

Whilst the latter part of all this spring migration is going on amongst birds that have either wintered with us, or only left our shores to seek more genial winter quarters in the immediate neighbourhood of the continent, the regular northern migration of birds from the far south, from the Mediterranean basin and Africa, begins; perhaps with the advent of such birds as Wheatears (*Saxicola œnanthe*), Chiffchaffs (*Phylloscopus rufus*), and Woodcocks (*Scolopax rusticola*), towards the end of March, although instances are on record where a few *avant courières* of the former species have been remarked in the north of England during the last week in February. With the advent of April, Migration becomes much more intense; as the month progresses species after species pours in from the South : delicate Warblers, Redstarts, the Cuckoo, the Wryneck, all the British Hirundines, Tree Pipits, and so on ; together with a steadily increasing stream of Waders, including the Common Sandpiper; also various species of Terns. Less intensity characterizes Migration of British individuals in May, although a vast number of birds are passing our islands during that period, on their way to the Arctic regions. During this month the last batch of our regular summer visitors make their appearance, which includes the Garden Warbler, the

Spotted Flycatcher, the Nightjar, the Swift, and
the Turtle Dove. As an instance of the important
stream of migrants still passing north, we learn, as
reported from the Eddystone Lighthouse, on the
last night of May, 1887 (cloudy with mist and
drizzling rain), that a Cuckoo was caught at the
lantern at 10 p.m.; that flocks of Sandpipers (some
individuals striking the light), Swifts, Swallows,
Wheatears, and Warblers continued to pass from
midnight till 3 a.m., or just in the gray dawn of
morning, great numbers of the last-mentioned birds
being killed against the lantern! By the first week
in June Spring Migration over the British Islands
has practically ceased, having prevailed to a more
or less extent for about four months.

Scarcely have the last spent pulses of the Great
Vernal Wave of Migration ceased to flow upon the
British Islands before the first ripples of the Autumn
tide begin to be apparent. By the middle of July
the Autumn Flight is inaugurated with the arrival
of the first Arctic Waders, either seen flying south
along our coasts or off them at sea, or heard even
at some distance inland crossing at great heights
over the star-lit sky. As we have already had
occasion to remark, a few old birds invariably
herald the rush of young. By the end of the
month and early in August the arrivals are more
numerous; young Knots and Gray Plovers, Common
Sandpipers, Lapwings, Ringed Plovers, Green-
shanks and Curlews. Small birds such as Swifts,
Wheatears, Willow Wrens, Wood Wrens, and

Whinchats, the young predominating, also begin
to pass during this period, doubtless from more
northern breeding-places. Right through August
the great wave of Migration is slowly gathering in
force, birds becoming more and more numerous on
our coasts, whilst our own summer birds are beginning to gather preparatory for flight in littoral
as well as in more inland districts. But little Migration takes place among our familiar summer birds
during August, however, notwithstanding the steady
influx of birds from other lands. Many of our
young Swifts and most of our old Cuckoos, however, move south in August: most of our other
summer migrants are now moulting ; all are songless, and much more skulking in their habits than
usual. Swallows and Martins begin to pack, and
very early in September a considerable number of
young birds depart south. During this month the
migration of various soft-billed or insectivorous
birds becomes much more intensified. Everywhere
Warblers, Swallows, Wheatears, Flycatchers, Ring
Ousels, Thrushes, Wagtails, and such-like species
are all speeding south, whilst the number of Wading
birds is very perceptibly increased. Especially is
this the case among Arctic species, such as Curlew
Sandpipers, Knots, Stints, Turnstones, Godwits,
Dunlins, Gray and Golden Plovers, a few Ducks
and Geese. Terns in flocks (principally young
birds) are also now moving south. At the beginning of the month the migrants are mostly
composed of young birds, especially among the

Passeres, but towards the end a perceptible pre-
ponderance of adult birds is noticed. At intervals
a general rush of one or two particular species will
be noted; one night it is Willow Wrens, White-
throats, Tree Pipits. and Sedge Warblers; another
night Thrushes, Wagtails, Flycatchers, and perhaps
large flocks of Plovers, Dunlins, and Swallows, with
an odd Woodcock or a Corn Crake, or a little party
of Robins and Meadow Pipits. No particular hour
seems chosen by each migrant: all are hurrying
along the great highways which are the common
property of each. By the beginning of October
the great majority of our summer birds are gone;
little more than the last lingering flocks of Swallows
remain behind. Many of the hardier birds continue
to pass our coasts, however, to more southern
latitudes, such as Buntings, Thrushes of various
kinds, Meadow Pipits, Larks, and Starlings; and
now the great autumn influx of birds coming to
the British Islands to winter continues day by day
to swell in volume. Waders and Water Birds still
continue to pass south, striking our coasts from the
north-east, and following the shore on their way, or
pass over more inland at great altitudes; others
arrive to winter upon them. By the end of October
the last of our own summer migrants have gone,
and our avifauna assumes quite a different character.

Up to the end of September the general stream
of Migration breaks upon our islands from the
North-east, then a very perceptible change of
direction takes place, and the predominating line of

Flight falls to nearly due East, or points south of East. This is the first sign of that gradually approaching Eastern Wave of Migration which trickles rather than flows until (on an average) the middle of the month. Then it suddenly assumes a more powerful flow, culminating in a grand and mighty influx of birds (young predominating), lasting almost incessantly for perhaps a fortnight : then a lull of a week or so occurs ; and then another grand wave of not quite the same magnitude and duration (adults predominating) breaks upon our eastern sea-board ; after which the Migrant Tide is more or less spent for the year, having drained the greater part of Western Asia and Eastern Europe of the majority of their hardiest non-insectivorous birds !

This westerly flowing tide of migrants is perhaps the most interesting Avian movement that takes place in our islands, because it is so enormous and so palpable to every eye. The number of species borne on its swelling stream is not very great, normally, but the number of individuals is almost past belief, and what is also to be remarked, they are all birds of exceptionally high powers of flight. Indeed this migration must be seen to be believed ; and a visit to our low-lying eastern coasts about the middle of October will rarely fail to convince the observer of its overpowering and bewildering vastness. Night and day the inrush of Migrants is constant and prodigious. For weeks I have repeatedly watched this marvellous Avian movement

on our eastern seaboard until almost bewildered by
the steady throbbing rush, rush, of arriving birds.
Thrushes, Larks, Goldcrests, Finches, Starlings,
Crows, Rooks, and Ring Doves comprise the bulk
of these migrants; but many other species arrive
in smaller numbers, and now and then a rare
straggler whose route has been abnormal. It is very
interesting to trace the approach and arrival of these
great Bird Waves, advancing from the continent
to our Islands. Here, for instance, is the movement of a very important wave from east to west
during the night of October 15—16th, 1885, traced
from Heligoland across the North Sea to the
British Islands. We find Gätke at his island
station of Heligoland making a note, referring to
the morning of the 15th, that the weather was
favourable for an important migration—Thrushes
and Woodcocks especially—the wind S.E., the
weather clear; but owing to the strong westerly
currents prevailing in the higher atmosphere, the
impending flight did not take place, or was not
visible from the island. In the evening the wind
rose and changed to E. by N., with clouds from
the S. and E., whilst in the night it backed to the
E. with thick clouds from the S.E. During the
day, however, Jackdaws and Hooded Crows, as well
as Thrushes, Pipits, Larks, Buntings, Jack Snipes
and Chaffinches were observed on passage, the
latter birds flying in thousands, but so high as to
be invisible, and only distinguished by their notes:
in the night the impending bird stream passed by,

and Gätke records an extraordinary strong migration of Larks, Thrushes, Starlings, Curlews and Plovers. On the 16th, in similar weather, "an extremely strong migration of Thrushes, Larks, Finches and Starlings" is recorded.

Here now are a few reports made on the same dates on the East coast of England, with the weather overcast and misty, and the wind E.N.E. From the Farne Islands we hear of a great rush of Fieldfares night and day, and of similar rushes at the Dudgeon Light-vessel off the Wash, 200 miles further south. We also hear of very large arrivals of Blackbirds by day and night striking the entire eastern coast-line of England from October 15th to the 18th; further, that Chaffinches arrived mainly in two large Flights between October 12—16th; that a very heavy immigration of Skylarks took place, the bulk of the birds arriving in enormous rushes on October 15th, 16th, and 17th; that enormous numbers of Starlings arrived from the 12th to the 19th, and large numbers were killed at the lights; that an almost continuous rush of Hooded Crows and Rooks appeared from the 15th to the 17th, between the Humber and Thanet.

Again, in the autumn of 1884, Gätke records on October 24th (clear and fine, with wind S.E.), " monstrous" numbers of Rooks, Hooded Crows and Jackdaws, the two latter species passing in mixed flights of ten and twelve minutes each, with short interruptions or gaps, the flocks extending as far as the eye could reach north and south from

9 a.m. to 1 p.m. A "succession of clouds" of Starlings also passed by overhead. On the East coast of England similar inrushes were remarked on days exactly corresponding. Again, in the same autumn we find that " immense rushes " of Starlings appeared on our eastern coasts during the latter half of October, by day and night. At Heligoland, Gätke remarks of this species : " enormous numbers " crossed between the 14th and 25th, especially on the 19th, " immense numbers "; on the 20th, " clouds of enormous numbers "; 21st, " astounding "; 22nd, " astounding flights, like clouds passing on." The autumn influx of Goldcrests is even more astonishing, when we bear in mind that this bird is the smallest Palæarctic species. From the Isle of May to the Channel Islands broad waves of this migrant strike our eastern and southern coasts, in varying numbers. In the autumn of 1882 this little bird reached Western Europe in marvellous numbers. Right through October they continued to arrive in enormous multitudes, two rushes being very pronounced, one on the night of the 7th and the morning of the 8th, the other on the night of the 12th and the morning of the 13th. At Heligoland, on the 28th and 29th, Gätke records : " a perfect storm of Goldcrests we have had—poor little souls ! —perching on the ledges of the window-panes of the lantern of our lighthouse, preening their feathers in the glare of the lamps; on the 29th, all the island swarmed with them, filling the gardens everywhere,

and over all the cliff—hundreds of thousands;
by 9 a.m. most of them had passed on again."
How many of these little pilgrims fell by the way
man may never know, but undoubtedly the great
majority never returned to the land of their birth!
The rush of Starlings and Sky Larks across the
North Sea in autumn especially is simply incredible.
For days and days together the Larks may be
watched coming into England in a scarcely broken
stream by day, and their warbling cries fill the air
at night as the great tide still flows on. Here is
Gätke's description of the migrating Starlings in
the autumn of 1883. On October 6th, "in
astounding flights, thousands upon thousands";
12th, "considerable numbers of astounding flights,
both overhead and in distance"; 13th, "still
passing, astounding numbers all day"; 26th,
"ditto, very high"; 27th, "night from 11 p.m.
myriads"; 28th, "immense." Correspondingly
large waves of these birds struck our coasts, and
might be described in similar terms.

We might thus go on dealing with species after
species in the same manner, but the limits of our
space forbid. Nevertheless these few intensely
interesting facts speak for themselves, vividly and
eloquently telling of that great Avian influx across
the wild North Sea. It is, however, worthy of
remark that in many cases vast waves of migrants
breaking over our islands do not touch Heligoland
at all, as is proved by the fact of enormous rushes
reaching our eastern coast-lines on days and nights

when little or no Flight has been remarked over the famous island. The same remarks apply to our own islands ; great waves and rushes being witnessed on the western or northern coast-lines, and scarcely any to correspond with them on the eastern, which clearly seems to demonstrate how wide and vast are these feathery tides, only breaking here and there on the coast, owing to local influences. The width of some of these vast east to west Bird Waves is enormous ; they have been known to break almost simultaneously on our shores from the Faroes or the Shetlands and the Orkneys in the north to the Channel Islands in the south, a distance of some 900 miles ; how much further they may have extended north and south remains a mystery ! Gätke, writing on these Avian Waves, on October 6th, 1883, with a N.E. wind and clear fresh weather, remarks : " Across the sea both sides of island (N. and S.), particularly on north side, countless numbers of *cornix* [Hooded Crows], *sturnus* [Starlings], and all kinds of small birds, all from E. to W. This occurrence happens not rarely ; during this ponderous migration there were on the island nearly no birds." Some of these Bird Waves are very persistent, for days and even weeks together, but normally the East to West autumn migration breaks principally upon our shores in two great floods, the first during the second and third weeks in October, and the second about a month later. The migration of each particular species in this great Avian wave varies considerably from year to year; sometimes it is completed in a few weeks, sometimes it extends over as many months. It is also worthy of remark how suddenly Migration will sometimes commence or cease. I have seen migrants coming for several days across the North Sea to the Wash—Larks especially— and for say three or four hours every morning, or all night, up to 10 or 11 a.m. the stream would flow, and then cease for the day. Hooded Crows often do this, migrating for days together in the morning. The first scattered flocks may be noted soon after sunrise; many alight on the mud-flats to feed or to rest, but the majority pass on to more inland districts, or set a southerly course along the coast-line. The irregularity of their appearance is sometimes remarkable. First perhaps an odd bird or so comes along; then in a few minutes a party of eight or ten; whilst at longer intervals occasional flocks appear, all flying in a slow, laboured, yet powerful manner, the individuals scattered about with no apparent approach to order.

The Mortality among these mighty bird hosts must be appalling, stupendous! The thousands of these pilgrims of the night that yearly kill themselves against the lighthouses and light-vessels round the British coasts represent a mere trifle in the general rate of mortality; the greater number perish at sea! In any case the sadly significant fact is only too apparent that only a fragment of these bird swarms—countless, one was going to say, as the sands on the shore—returns in spring.

Where they go and what their ultimate fate is a matter shrouded in mystery! It teaches us, however, to realize the fecundity and the mortality pertaining to Avian Life.

The origin of this great East to West Migration Wave must undoubtedly be sought in long-past ages. That the movement is a deeply-rooted one is evident from the persistence with which it is undertaken, even by some few species of comparatively weakly flight. Doubtless it originated from a prevailing low winter temperature in Eastern Europe and Western Asia, and a comparatively high winter temperature in Western Europe owing to Gulf Stream influence. Had it not been for the latter current this great Wave would undoubtedly have drifted South instead of West. That this difference of temperature exists is a fact no one will attempt to gainsay. The winters of South Russia and South-west Siberia, in the same latitudes as the South of France, are as remarkable for their normal severity as the winters of West Europe within the same parallels of latitude are for their normal mildness. That this is due to the warm ocean current that flows along the coasts of the latter is also an undoubted fact. There can be no doubt whatever (after what we have already learnt of Migration) that this difference of climate has had a marked effect on Avian life, and been the initiating cause of this East to West Migration Flight. Now in remote yet Post-Glacial Ages, the British Islands, the entire German Ocean, the English

Channel, and parts of the Bay of Biscay formed one land mass connecting with Scandinavia by way of the Shetlands, and extending south-west round the south of Ireland to the south-west of France. Migrants then reached this portion of Europe without having any sea-flight at all ; it was simply an overland journey to the mild coast regions of continental land—the vast bulging west peninsula of Europe. But great submergence has taken place ; the North Sea, the English Channel, the Irish Sea have been formed, and great areas of land have sunk along the Atlantic sea-board, leaving matters as we now see them. The stream of Migration, however, continued—the custom of visiting these mild western regions was too deeply rooted a custom to be relinquished ; and not even the widening area of gradually accumulating sea has succeeded in stamping it out. In those far-off days the difference of climate was evidently even more acute than it is now, as is proved not only by geological evidence, but by the still surviving relics of a flora in the south of England and Ireland that belongs decidedly to Italy and the South, rather than to Scandinavia! We have seen in earlier portions of this work how birds still continue to follow submerged routes ; across that ancient Lemuria, for instance, now lying beneath Indian Ocean waves ; we have in this East to West stream of Migration another instance, although on a smaller scale, and consequently followed by many more species than travel by that submerged route

between India and South Africa! We might here
also allude to the interesting fact of several birds
being regular summer migrants to France and
Holland, and yet never visiting the British Islands,
except perhaps as stragglers on abnormal flight.
The Great Reed Warbler (*Acrocephalus turdoides*)
and the Blue-headed Wagtail (*Motacilla flava*) may
be cited as particularly good instances of birds
common enough across the English Channel, yet
as rare in our islands.   I am of opinion that these
birds did not extend their emigrations or migrations
so far north in Western Europe before our islands
became detached from continental land ; they are
more recent emigrants which found the English
Channel a barrier to any further western extension.

   There are one or two other matters to which we
must allude ere dismissing the present subject.
First, it may be as well to glance at the Cross
Migration taking place in autumn, just as we
remarked it in spring, only of course the direction
is exactly reversed.   A very interesting instance of
this peculiar Flight is reported on the 17th of
October, 1885, from the Isle of May Lighthouse in
the Firth of Forth.   We cannot do better than let
Mr. Agnew the light-keeper speak for himself.
"A great rush of migrants at Isle of May.   The
Woodcock was killed on the S.E. side, and
the other birds on the N.W. side.   Thus the
migrants have been exactly meeting one another ;
and this is just as I would have expected—the
Woodcocks *coming* to us, and the others *leaving*

us." Now this is an intensely interesting fact, and confirms very strongly our opinion above expressed, that in some cases at least the birds flying south are merely passing our coasts on passage from North-Western Europe; for among the species that struck on the N.W. side in this instance was the Redwing, clearly demonstrating that in this case the northern stream was on its normal way south *viâ* the British coasts, and was not composed of our own birds leaving us to winter further south. Whinchats and Willow Wrens were also in the stream—species the individuals of which breed in Scotland normally leave that district, say in September. A similar Cross Migration is also apparent during autumn in the south-western portions of the North Sea.

We have also the same facts presented to us of Migrants waiting for favourable winds; and on the south coast of England the same double streams of birds, one passing nearly due east to the Straits of Dover, composed of various kinds of summer migrants on their way south, passed by another stream of hardier birds coming into our islands for the winter, and passing as nearly due west.

It remains for us now to notice the various Winter Migrations passing from the continent to our islands, or Local Movements among the birds wintering with us. Changes of temperature are the great initiating causes of these minor waves of Nomadic Migration, and some of the rushes which then occur are both interesting and important. Thus severe weather in

the northern districts will send thousands of Larks,
Snow Buntings, Finches, Crows, and Starlings
drifting along the coasts towards the south. At
the Galloper Light-vessel, stationed off the mouth
of the Thames, we have a very interesting instance
of this Nomadic Migration, taken from scores of
similar instances. On the nights of January 21st
to 23rd, 1885, no less than 226 birds were killed,
all refugees from a severe spell of winter in more
northern districts. We have many instances where
our eastern coast districts have swarmed with Snow
Buntings and other hard-billed birds, and of spasmodic arrivals of others from the continent during
exceptionally hard weather, which will, as any
observer may remark, clear a district of birds very
quickly for the time. In fact, all the winter
through Nomadic Migration is in progress on no
uncertain scale. Birds are constantly passing to
and fro, now sparingly, now in bewildering numbers;
from every point the same story comes, proving
absolutely that at least some species are actually in
a constant state of passage through every month of
the twelve. In Larks and Starlings this fact is
specially noticeable. Some birds are certainly more
highly susceptible to a change of weather than
others, especially those that obtain their food from
the ground. Lapwings, Larks,[1] and Snow Bunt-

---

[1] It may be for this reason that Larks always prefer a
winter haunt on the highest ground in the district they frequent,
where a coming change of temperature is probably more readily
and quickly detected than in the valleys. My invariable ex-

ings will unerringly foretell a snow-storm hours in advance, and commence a nomadic migration to escape its dangers. From all this constant Passage one vastly important fact may be deducted. It shows us how the individuals of a species are kept well intermixed, and thus furnish those all-necessary facilities for interbreeding that are the dominating influences in preserving those species, and keeping them up to their normal standards of constancy.

We would fain have entered more fully into the Migration of British birds, but we are compelled with reluctance to confine our chapter to the limits it has now reached. It would have been a very easy matter to fill our volume with the stirring story of their movements alone; but we trust that sufficient has been said to make the whole subject of Migration reasonably and tolerably complete, notwithstanding its unavoidable meagreness. The following tables will give the reader some idea of the general movements of birds in the British Islands.

The student will also find appended Blank Tables for the next eight years in which he may record, in his own particular district, the Migration Flight of certain common or fairly well distributed species whose movements in Spring and Autumn may be remarked by the most casual observer of natural phenomena.

perience of the Sky Lark (*Alauda arvensis*), is that the bird always shuns a valley, and frequents the most elevated pastures during winter.

TABLE I. SHOWING THE DURATION OF FLIGHT OF SPRING MIGRANTS.

SPECIES.	JAN.	FEB.	MAR.	APR.	MAY.	JUN.	JULY.	AUG.	SEPT.	OCT.	NOV.	DEC.
		ARRIVALS.							DEPARTURES.			
Hobby (*Falco subbuteo*).	++++	++++	++ ×	× ×	× × × ×	...	...	...	× × ×	× × ×	++++ ×	++++
Kestrel (*Falco tinnunculus*).												
Osprey (*Pandion haliaëtus*).												
Honey Buzzard (*Pernis apivorus*).												
Hen Harrier (*Circus cyaneus*).												
Song Thrush (*Turdus musicus*).												
Ring Ouzel (*Merula torquata*).												
Nightingale (*Erithacus luscinia*).												
Whinchat (*Pratincola rubetra*).												
Redstart (*Ruticilla phoenicurus*).												
Wheatear (*Saxicola oenanthe*).												
Spotted Flycatcher (*Muscicapa grisola*).												
Pied Flycatcher (*Muscicapa atricapilla*).												
Grasshopper Warbler (*Locustella locustella*).												
Sedge Warbler (*Acrocephalus phragmitis*).												
Reed Warbler (*Acrocephalus arundinaceus*).												
Marsh Warbler (*Acrocephalus palustris*).												
Blackcap (*Sylvia atricapilla*).												
Garden Warbler (*Sylvia hortensis*).												
Whitethroat (*Sylvia cinerea*).												
Lesser Whitethroat (*Sylvia curruca*).												
Wood Wren (*Phylloscopus sibilatrix*).												
Willow Wren (*Phylloscopus trochilus*).												
Chiffchaff (*Phylloscopus rufus*).												
Red-backed Shrike (*Lanius collurio*).												
Swallow (*Hirundo rustica*).												
House Martin (*Chelidon urbica*).												
Sand Martin (*Cotyle riparia*).												

Pied Wagtail (*Motacilla yarrellii*).
White Wagtail (*Motacilla alba*). ...
Blue-headed Wagtail (*Motacilla flava*). ...
Yellow Wagtail (*Motacilla raii*). ...
Tree Pipit (*Anthus arboreus*). ...
Swift (*Cypselus apus*). ...
Nightjar (*Caprimulgus europæus*). ...
Hoopoe (*Upupa epops*). ...
Wryneck (*Jynx torquilla*). ...
Cuckoo (*Cuculus canorus*). ...
Turtle Dove (*Turtur auritus*). ...
Quail (*Coturnix communis*). ...
Corn Crake (*Crex pratensis*). ...
Spotted Crake (*Crex porzana*). ...
Stone-Curlew (*Œdicnemus crepitans*). ...
Kentish Plover (*Ægialophilus cantianus*). ...
Dotterel (*Eudromias morinellus*). ...
Rednecked Phalarope (*Phalaropus hyperboreus*)
Whimbrel (*Numenius phæopus*). ...
Ruff (*Machetes pugnax*). ...
Common Sandpiper (*Totanus hypoleucus*). ...
Greenshank (*Totanus glottis*). ...
Black-tailed Godwit (*Limosa melanurus*). ...
Dunlin (*Tringa alpina*). ...
Woodcock (*Scolopax rusticola*). ...
Sandwich Tern (*Sterna cantiaca*). ...
Arctic Tern (*Sterna arctica*). ...
Common Tern (*Sterna hirundo*). ...
Lesser Tern (*Sterna minuta*). ...
Garganey (*Anas circia*). ...

NOTE.—The number of crosses in each month in this and the following table indicates the intensity of the Migration. † Denotes abnormal appearances. ‡ Varying numbers of individuals habitually remain to winter on our shores.

TABLE II. SHOWING THE DURATION OF FLIGHT OF AUTUMN MIGRANTS.

SPECIES.	ARRIVALS.						DEPARTURES.					
	JULY.	AUG.	SEPT.	OCT.	NOV.	DEC.	JAN.	FEB.	MAR.	APR.	MAY.	JUNE.
Short-eared Owl (*Strix brachyotus*).		×	×	×	×	⋮	⋮	⋮	×	×	×	
Missel Thrush (*Turdus viscivorus*).		×	×	×	×	⋮	⋮	×	×	×	×	
Redwing (*Turdus iliacus*).		×	×	×	×	×	⋮	×	×	×		
Song Thrush (*Turdus musicus*).			×	×	×	⋮	×	×	×	×	×	
Fieldfare (*Turdus pilaris*).			×	×	×	×	×	×	×	×	×	
Blackbird (*Merula merula*).			×	×	×	×	×	×	×			
Robin (*Erithacus rubecula*).			×	×	×		×	×	×	×	×	
Black Redstart (*Ruticilla tithys*).			×	×	×		×	×	×	×	×	
Goldcrest (*Regulus cristatus*).			×	×	×		⋮	⋮	×	×	×	
Hedge Accentor (*Accentor modularis*).				×	×	×	⋮	⋮	×	×	×	×
Wren (*Troglodytes parvulus*).	+			×	×	⋮	⋮	⋮	×	×	×	
Carrion Crow (*Corvus corone*).		×	×	×	×	×	⋮	×	×	—	×	
Hooded Crow (*Corvus cornix*).		×	×	×	×	×	×	×	×	×	×	
Rook (*Corvus frugilegus*).		⋮	×	×	×	×	×	×	×	×	×	
Jackdaw (*Corvus monedula*).	×	×	×	×	×	×	×	×	×	×	×	
Pallas's Gray Shrike (*Lanius major*).										×		
Great Gray Shrike (*Lanius excubitor*).							×	×	×	×	×	
Starling (*Sturnus vulgaris*).			×	×	×	×	×	×	×	×	×	
House Sparrow (*Passer domesticus*).	×		×	×	×	×	⋮	⋮	×	×	⋮	
Tree Sparrow (*Passer montanus*).	×	×	×	×	×	×	×	×	×	×	×	
Greenfinch (*Fringilla chloris*).			×	×	×	×	×	×	×	×	×	
Goldfinch (*Fringilla carduelis*).			⋮	×		⋮	⋮	×	×	×	×	
Brambling (*Fringilla montifringilla*).			×	×	×	⋮	×	×	×	×	×	
Chaffinch (*Fringilla cœlebs*).		×	×	×	×	⋮	×	×	×	×	×	
Siskin (*Fringilla spinus*).		×	×	×	×	×	⋮	⋮	×	×	×	
Linnet (*Linota cannabina*).	×	×	×	×	×	×	×	×	×	×	×	×

Twite (*Linota flavirostris*).
Mealy Redpole (*Linota linaria*).
Hawfinch (*Coccothraustes vulgaris*).
Snow Bunting (*Emberiza nivalis*).
Corn Bunting (*Emberiza miliaria*).
Yellow Bunting (*Emberiza citrinella*).
Reed Bunting (*Emberiza schœniclus*).
Meadow Pipit (*Anthus pratensis*).
Skylark (*Alauda arvensis*).
Kingfisher (*Alcedo ispida*).
Ring Dove (*Columba palumbus*).
Stock Dove (*Columba œnas*).
Turnstone (*Strepsilas interpres*).
Golden Plover (*Charadrius pluvialis*).
Gray Plover (*Charadrius helveticus*).
Lapwing (*Vanellus cristatus*).
Curlew (*Numenius arquatus*).
Whimbrel (*Numenius phæopus*).
Redshank (*Totanus calidris*).
Bar-tailed Godwit (*Limosa rufa*).
Knot (*Tringa canutus*).
Dunlin (*Tringa alpina*).
Purple Sandpiper (*Tringa maritima*).
Sanderling (*Calidris arenaria*).
Woodcock (*Scolopax rusticola*).
Common Snipe (*Scolopax gallinago*).
Jack Snipe (*Scolopax gallinula*).
Hooper (*Cygnus musicus*).

TABLE II. SHOWING THE DURATION OF FLIGHT OF AUTUMN MIGRANTS.

SPECIES.	ARRIVALS.						DEPARTURES.					
	JULY.	AUG.	SEPT.	OCT.	NOV.	DEC.	JAN.	FEB.	MAR.	APR.	MAY.	JUN.
Bewick's Swan (*Cygnus bewicki*).				×	×	× ×	× ×		×	×	×	
Bean Goose (*Anser segetum*).				× ×	× ×				×	× ×		
Pink-footed Goose (*Anser brachyrhynchus*).				× ×	× ×				× ×	× ×	× ×	×
Gray-lag Goose (*Anser cinereus*).				×	× ×	×		×	× ×	× ×	× ×	
White-fronted Goose (*Anser albifrons*).			× ×	× ×	× ×				×	× ×	× ×	×
Lesser White-fronted Goose (*Anser minutus*).				×						×		
Brent Goose (*Bernicla brenta*).				×	× ×	×		×	× ×	× ×	× ×	
White-bellied Brent Goose (*Bernicla glaucogaster*)				×	×	×	×	×	× ×	× ×	×	
Bernacle Goose (*Bernicla leucopsis*).				×	× ×	×		×	× ×	× ×	× ×	× ×
Gadwall (*Anas strepera*).			×	× ×	× ×	×			× ×	× ×	×	
Pintail (*Anas acuta*).			× ×	× ×	× ×	×			× ×	× ×	×	
Wigeon (*Anas penelope*).			× ×	× ×	× ×	×		×	× ×	× ×	×	
Teal (*Anas crecca*).			× ×	× ×	× ×	×		×	× ×	× ×	×†	
Shoveller (*Anas clypeata*).			×	× ×	× ×				× ×	× ×	×	
Mallard (*Anas boschas*).				× ×	× ×					× ×		
Pochard (*Fuligula ferina*).			×	× ×	× ×	×		×	× ×	× ×	×	
Scaup (*Fuligula marila*).				× ×	× ×	×	×	× ×	× ×	× ×	×	
Tufted Duck (*Fuligula cristata*).			×	× ×	× ×	×		× ×	× ×	× ×	× ×	
Golden-eye (*Clangula clangula*).				× ×	× ×	×		×	× ×	× ×	×	
Long-tailed Duck (*Harelda glacialis*).				×	× ×	×	×	×	×	×	× ×	
Common Scoter (*Œdemia nigra*).	×	×	× ×	× ×	×	×			×	×	×	
Velvet Scoter (*Œdemia fusca*).		×	× ×	×	×	×			×	× ×	× ×	
Goosander (*Mergus merganser*).		×	×	× ×	× ×	×		×	× ×	× ×	× ×	

NOTE.—The majority of these species breed in the British Isles, but are either increased in numbers or replaced by migratory individuals in autumn. † Denotes abnormal appearances.

TABLE III. SHOWING THE DURATION OF FLIGHT OF COASTING MIGRANTS.

SPECIES.	JAN.	FEB.	MAR.	APR.	MAY	JUN.	JULY	AUG.	SEPT.	OCT.	NOV.	DEC.
Rough-legged Buzzard (*Archibuteo lagopus*).		×+×	×××	××	××	..	..	..		××	××	++
Pied Wagtail (*Motacilla yarrellii*).	+	+	×××	××	××	..	×	×	×	××	×+	++
Gray Wagtail (*Motacilla sulphurea*).				××	××	..		×	××	×××	+×	+
Rock Pipit (*Anthus obscurus*).	+	+	+××	××	××	×		××	××	××	××	—
Spoonbill (*Platalea leucorodia*).					..	..	..					
Little Crake (*Crex parva*).				×	××			×	××	+×	+×	×+
Crane (*Grus cinerea*).					××		×	××	×××	××	××	
Ringed Plover (*Ægialitis hiaticula*).	+	+×+	×	××	×××	..	×.	××	×××	××	×	
Dotterel (*Eudromias morinellus*).			:+	××	×××	×	×	××	×××	××	×	×+
Gray Plover (*Charadrius helveticus*).		+×	×	×	×××	×		×	××	×	×	
Avocet (*Recurvirostra avocetta*).					×	×			×××			
Whimbrel (*Numenius phaeopus*).	+	+	×	××	××	×	×	××	×××	×+	+×	+
Ruff (*Machetes pugnax*).				×	×××	×	×	××	×××	×+	×+	—
Green Sandpiper (*Totanus ochropus*).				×	××	×	×	××	×××	×		
Wood Sandpiper (*Totanus glareola*).					×	×	×.	×	××	×	×	×+
Redshank (*Totanus calidris*).		++?+	×+?×	××	××××	×	××	××	×××	××	++×	+×+
Dusky Redshank (*Totanus fuscus*).				×	×××	×	×	×	×××	××	××	—
Greenshank (*Totanus glottis*).	+	+	+	××	××	×	×	××	×××	××	+×	+
Knot (*Tringa canutus*).				××	××		×	××	×××	×	×	+
Curlew Sandpiper (*Tringa subarquata*).		++?+	×+?×	××	××	×	×	××	×××	××	+×	++?
Dunlin (*Tringa alpina*).	++	+×+	+?×	××	××	×	×	××	×××	××	?+	?+
Little Stint (*Tringa minuta*).			×	×	×		×	××	×××	××	+	+
Temminck's Stint (*Tringa temmincki*).				×××	××	..		×	×××	××		
Sanderling (*Calidris arenaria*).	++	+×	+	×××	×××	×	×	×	×××	×××	××	+
Woodcock (*Scolopax rusticola*).				××	×	..		×	×	×××	××	×+

(Table columns at left — JAN. FEB. MAR. APR. MAY JUN. — are headed **NORTHWARDS.**; columns at right — JULY JUL./AUG. SEPT. OCT. NOV. DEC. — are headed **SOUTHWARDS.**)

Black Tern (*Sterna nigra*).

Great Skua (*Stercorarius catarrhactes*).

Richardson's Skua (*Stercorarius richardsoni*).

Pomarine Skua (*Stercorarius pomarinus*).

Black-necked Grebe (*Podiceps nigricollis*).

Gray-lag Goose (*Anser cinereus*).

Bean Goose (*Anser segetum*).

White-fronted Goose (*Anser albifrons*).

Lesser White-fronted Goose (*Anser minutus*).

Brent Goose (*Bernicla brenta*).

White-bellied Brent Goose (*Bernicla glaucogaster*).

Bernacle Goose (*Bernicla leucopsis*).

Gadwall (*Anas strepera*).

Pintail (*Anas acuta*).

Wigeon (*Anas penelope*).

Teal (*Anas crecca*).

Shoveller (*Anas clypeata*).

Mallard (*Anas boschas*).

Pochard (*Fuligula ferina*).

Scaup (*Fuligula marila*).

Tufted Duck (*Fuligula cristata*).

Golden-eye (*Clangula clangula*).

Long-tailed Duck (*Harelda glacialis*).

Common Scoter (*Œdemia nigra*).

Velvet Scoter (*Œdemia fusca*).

Goosander (*Mergus merganser*).

Red-breasted Merganser (*Mergus serrator*).

NOTE.—The number of crosses indicates the intensity of Migration. Where an obelisk is inserted it denotes that a certain number of individuals remain in our islands during the winter, or are resident with us.

T

TABLE IV. SHOWING THE PRINCIPAL MOVEMENTS OF NOMADIC MIGRANTS.

SPECIES.	AUTUMN (Southwards)						WINTER AND SPRING (Northwards)					
	JULY.	AUG.	SEPT.	OCT.	NOV.	DEC. (WINTER)	JAN.	FEB.	MAR.	APR.	MAY.	JUNE.
Hawk Owl (*Surnia funerea*).				×	×	:	×					
Tengmalm's Owl (*Noctua tengmalmi*).				×	×	×		×				
Snowy Owl (*Surnia nyctea*).				×	×	×	:	:	:			
Eagle Owl (*Bubo maximus*).				×	×	:	:	:	:	×	×	
Great Titmouse (*Parus major*).	×	×	× × ×	× × × ×	× × ×	:	:	:	× × × ×	× × × ×	× × × ×	
Blue Titmouse (*Parus caeruleus*).	×	×	× × ×	× × × ×	× × ×	:	:	:	× × × ×	× × × ×	× × × ×	
Coal Titmouse (*Parus ater*).	×	×	× × ×	× × × ×	: : × :	:	:	:	× × × ×	× × × ×	× × × ×	
Long-tailed Titmouse (*Acredula caudata*).	×	×	×	× × × ×	:	:	:	:	× × × ×	× × × ×	× × × ×	
Nutcracker (*Nucifraga caryocatactes*).						:	:	:				
Waxwing (*Ampelis garrulus*).	×	:	:	× ×	× ×	× ×	× ×	×	:	:	:	
Common Crossbill (*Loxia curvirostra*).			× ×	× ×	×	×	×	×	×	×	×	×
Parrot Crossbill (*Loxia pityopsittacus*).						:	:	:	:	:	:	:
White-winged Crossbill (*Loxia bifasciata*).						×	:	:	:	:	:	:
American ,, (*Loxia leucoptera*).						:	:	:	:	:	:	:
Pine Grosbeak (*Pinicola enucleator*).						×	:	:	:	:	:	:
Lapland Bunting (*Emberiza lapponica*).		×	×	× ×	× ×	×	×	× ×	×	× ×		
Shore Lark (*Otocoris alpestris*).	×	×	×	× × ×	× ×	× ×	×	× ×	×	× ×		
Heron (*Ardea cinerea*).		×	×	× × ×	× ×	:	×	× ×	× ×	× ×		
Bittern (*Botaurus stellaris*).	×	×	×	× × ×	× ×	× ×	: ×	× ×				

Species						
Great Bustard (*Otis tarda*).			×	×	×	
Little Bustard (*Otis tetrax*).			×	×	×	
Gray Phalarope (*Phalaropus fulicarius*).						
Little Gull (*Larus minutus*).	×		×	×	×	
Sabine's Gull (*Xema sabinii*).	×		×	×	×	
Ross's Gull (*Rhodostethia rosii*).	×		×	×	×	
Iceland Gull (*Glaucus leucopterus*).	×	×	×	×	×	
Ivory Gull (*Pagophila eburnea*).	×	×	×	×	×	
Buffon's Skua (*Stercorarius buffoni*).			×	×		
Little Auk (*Mergulus alle*).	×	×	×	×	×	
Great Northern Diver (*Colymbus glacialis*).	×	×	×	×	×	
White-billed Diver (*Colymbus adamsi*).	×	×	×	×		
Black-throated Diver (*Colymbus arcticus*).	×	×	×	×	×	
Red-throated Diver (*Colymbus septentrionalis*).	×	×	×	×	×	
Great Crested Grebe (*Podiceps cristatus*).	×	×	×			
Red-necked Grebe (*Podiceps rubricollis*).	×	×	×			
Sclavonian Grebe (*Podiceps cornutus*).	×	×	×	×		
Harlequin Duck (*Clangula histrionica*).			×	×	×	
Surf Scoter (*Œdemia perspicillata*).	×	×	×	×	×	
Steller's Eider (*Somateria stelleri*).	×	×	×	×	×	
King Eider (*Somateria spectabilis*).	×	×	×	×	×	
Smew (*Mergus albellus*).	×	×	×	×	×	

NOTE.—The number of crosses indicates the frequency or rarity of appearances.

TABLE V. SHOWING THE MOVEMENTS OF VERTICAL MIGRANTS.

SPECIES.	ASCENDING.						DESCENDING.					
	JAN. FEB.	MAR.	APR.	MAY.	JUNE.	JULY.	AUG.	SEPT.	OCT.	NOV.	DEC.
Merlin (*Falco esalon*). ...		× ×	× ×	...	...	...	×	×	× ×		
Stonechat (*Pratincola rubicola*).	× ×	× ×	× ×	...	...	...	...	× × ×	× ×		
Linnet (*Linota cannabina*).		× ×	× ×	× ×	...	...	× ×	× × ×	× × ×		
Twite (*Linota flavirostris*). ...			× × ×	× ×	...	...	× ×	× × ×	× ×		
Lesser Redpole (*Linota rufescens*).		× ×	× ×	× ×	...	...	× ×	× ×	× ×		
Pied Wagtail (*Motacilla yarrellii*).		× ×	×	...	...	...		×	× ×		
Gray Wagtail (*Motacilla sulphurea*).			×	×	...	...		× ×	× ×		
Meadow Pipit (*Anthus pratensis*). ...		× ×	×	...	...	...	...	× × ×	× × ×		
Wood Lark (*Alauda arborea*).		× ×		...	...	...	...	× ×	× ×		
Sky Lark (*Alauda arvensis*).	×	× ×	× ×	×	...	...	...	× ×	× ×	×	
Golden Plover (*Charadrius pluvialis*). ...	× ×	× ×	× ×	...	...	...	...	× × ×	× × ×	×	
Lapwing (*Vanellus cristatus*).	×	× ×	× ×	×	...	...	× ×	× ×	× ×	×	
Curlew (*Numenius arquatus*).	×	× ×	× ×	×	...	...	× ×	× ×	× ×	×	
Dunlin (*Tringa alpina*). ...		× ×	× ×	...	...	...	× ×	× ×	× ×	×	

NOTE.—The number of crosses indicates the intensity of Migration.

TABLE VI. SHOWING THE PRINCIPAL OCCURRENCES OF ABNORMAL MIGRANTS.

SPECIES.	JAN	FEB	MAR	APR (SPRING)	MAY	JUNE	JULY	AUG	SEPT	OCT	NOV	DEC
Orange-legged Hobby (*Falco vespertinus*).	×				×				×	×		
Brown Jer-Falcon (*Falco gyrfalco*)....			×									WINTER
Iceland Jer-Falcon (*Falco gyrfalco candicans*).			×		×							WINTER
White Jer-Falcon (*Falco candicans*).												WINTER
Lesser Kestrel (*Falco cenchris*). ...											×	
Swallow-tailed Kite (*Elanoïdes furcatus*).	×				×				×			
Black Kite (*Milvus ater*). ...										×	×	×
Lesser Spotted Eagle (*Aquila nævia*).												
American Goshawk (*Astur atricapillus*).											× ·	×
Little Owl (*Noctua noctua*). *	×										×	×
American Hawk Owl (*Surnia hudsonia*).										×	×	WINTER
Scops Owl (*Scops scops*).					×					×	×	×
White's Thrush (*Geocichla varia*). ...												WINTER
Siberian Thrush (*Geocichla sibirica*).										×		WINTER
Black-throated Ouzel (*Merula atrigularis*).										×		
Black-bellied Dipper (*Cinclus melanogaster*).		×			×				×	×		
Arctic Blue-throat (*Erithacus suecica*).	×	×			×				×	× ×		
Rock Thrush (*Monticola saxatilis*).		×				×	×		×	×		
Desert Chat (*Saxicola deserti*). ...						×						
Black-throated Chat (*Saxicola stapazina*).	×	×	×		×					×		×
Isabelline Chat (*Saxicola isabellina*).		×										
Red-breasted Flycatcher (*Muscicapa parva*).	×								×	×	× ×	
White-collared Flycatcher (*Muscicapa collaris*).												
Aquatic Warbler (*Acrocephalus aquaticus*).					×				×	×		
Great Reed Warbler (*Acrocephalus turdoides*).*						×			×			
Icterine Warbler (*Hypolais hypolais*).					×	×			×			
Barred Warbler (*Sylvia nisoria*). ...									×			
Orphean Warbler (*Sylvia orphea*).		×					×	×	×			
Rufous Warbler (*Sylvia galactodes*).					×	×			×	×		

Column headers (rotated): WINTER … AUTUMN … AUTUMN … AUTUMN … AUTUMN

Species						
Yellow-browed Warbler (*Phylloscopus superciliosus*).						
Firecrest (*Regulus ignicapillus*).						
Ruby-crowned Wren (*Regulus calendula*).*						
Alpine Accentor (*Accentor alpinus*).						
Wall Creeper (*Tichodroma muraria*).						
Golden Oriole (*Oriolus galbula*).						
Gold-vented Thrush (*Pycnonotus capensis*).						
Lesser Gray Shrike (*Lanius minor*).						
Woodchat Shrike (*Lanius rufus*).						
Rose-coloured Pastor (*Pastor roseus*).						
Red-winged Starling (*Agelaeus phoeniceus*).						
Rusty Grakle (*Scolecophagus ferrugineus*).						
Meadow Starling (*Sturnella magna*).						
Scarlet Rose Finch (*Carpodacus erythrinus*).						
Canary (*Fringilla canaria*).						
Serin (*Fringilla serinus*).						
Greenland Redpole (*Linota hornemanni*).						
White-throated Sparrow (*Zonotrichia albicollis*).						
Brandt's Siberian Bunting (*Emberiza cioides*).						
Rustic Bunting (*Emberiza rustica*).						
Little Bunting (*Emberiza pusilla*).						
Ortolan Bunting (*Emberiza hortulana*).						
Black-headed ,, (*Emberiza melanocephala*).						
White-bellied Swallow (*Tachycineta bicolor*).*						
Purple Martin (*Progne purpurea*).						
Red-throated Pipit (*Anthus cervinus*).						
Richards' Pipit (*Anthus richardi*).						
Tawny Pipit (*Anthus campestris*).						
Alpine Pipit (*Anthus spinoletta*).						
Crested Lark (*Alauda cristata*).						
Short-toed Lark (*Alauda brachydactyla*).						

NOTE.—An * denotes that the month of capture is unknown.

TABLE VI. SHOWING THE PRINCIPAL OCCURRENCES OF ABNORMAL MIGRANTS.

SPECIES.	JAN.	FEB.	MAR.	APRIL	MAY	JUNE	JULY	AUG.	SEPT.	OCT.	NOV.	DEC.
White-winged Lark (*Alauda sibirica*).						×	...	...	...	...	×	
Calandra Lark (*Alauda calandra*).*					× ×	×	×	×	×	×	AUTUMN	
Alpine Swift (*Cypselus melba*).						×	×					
Needle-tailed Swift (*Chaetura caudacuta*).							×					
Isabelline Nightjar (*Caprimulgus aegyptius*).							...	...	...	×	×	
Red-necked Nightjar (*Caprimulgus ruficollis*).							...	...	...	×		
Bee-eater (*Merops apiaster*).			×	×	× ×	×	...	...	...	×		
Roller (*Coracias garrula*).			× ×	×			...	...	...	×		
Indian Roller (*Coracias indicus*).			...				...	...	...	×	×	
Belted Kingfisher (*Ceryle alcyon*).			...		×		...	...	...	×		
Great Spotted Cuckoo (*Cuculus glandarius*).				×	×	...	×	×	×	× ×		
Yellow-billed Cuckoo (*Coccyzus americanus*).										× ×		×
Black-billed „ (*Coccyzus erythrophthalmus*).									×			
Passenger Pigeon (*Ectopistes migratorius*).						× ×	×		...	× × × ×		
Eastern Turtle Dove (*Turtur orientalis*).						×			×			
Pallas's Sand Grouse (*Syrrhaptes paradoxus*).						×	...	×	×		×	
Purple Heron (*Ardea purpurea*).					× ×	×	×	×	×	×	×	
Great White Egret (*Ardea alba*).						×	...	...	×	×		
Little Egret (*Ardea garzetta*).					... ×	×		...		×		
Squacco Heron (*Ardea comata*).					... ×	...	SUMMER			...	×	
Buff-backed Heron (*Ardea bubulcus*).									AUTUMN	×		
Night Heron (*Nycticorax nycticorax*).			×		×	×	...	...	...	×		
Little Green Heron (*Butorides virescens*).									...	×		
American Bittern (*Botaurus lentiginosus*).	×								...	×	×	
Little Bittern (*Botaurus minutus*).		×		...	...	...			...	×		
Glossy Ibis (*Ibis falcinellus*).		...		SPRING					×	×	×	×
White Stork (*Ciconia alba*).			×	×	×						AUTUMN	×

NOTE.—An * denotes that the month of capture is unknown.

Species	SPRING			AUTUMN	AUTUMN (rarely)
Black Stork (*Ciconia nigra*).		× ×		×	:
Demoiselle Crane (*Grus virgo*).		×		:	:
Macqueen's Bustard (*Otis macqueeni*).	:	×		:	:
Little Ringed Plover (*Ægialitis minor*).	:	:		×	:
Killdeer Plover (*Ægialitis vocifera*).	×		×	:	×
Caspian Sand Plover (*Ægialophilus asiaticus*).				:	:
Sociable Lapwing (*Vanellus gregarius*).				:	:
Asiatic Golden Plover (*Charadrius fulvus*).				× ×	:
American Golden ,, (*Charadrius virginicus*).	:	:		× × ×	:
Cream-coloured Courser (*Cursorius gallicus*).		×		×	×
Common Pratincole (*Glareola pratincola*).	:	× ×	×	× × ×	×
Common Stilt (*Himantopus melanopterus*).	×	×		× ×	×
Esquimaux Curlew (*Numenius borealis*).		×		× × × × × ×	×
Bartram's Sandpiper (*Totanus bartrami*).				× × ×	:
Spotted Sandpiper (*Totanus macularius*).				× × × ×	×
Solitary Sandpiper (*Totanus solitarius*).	:			: :	:
Yellowshank (*Totanus flavipes*).				× × ×	×
Red-breasted Snipe (*Ereunetes griseus*).				× ×	:
Bonaparte's Sandpiper (*Tringa bonapartei*).	: ×	× ×	×	: ×	:
Broad-billed Sandpiper (*Tringa platyrhyncha*).				× × ×	×
Pectoral Sandpiper (*Tringa pectoralis*).				× × ×	×
American Stint (*Tringa minutilla*).	:	× × ×	:	× × ×	:
Buff-breasted Sandpiper (*Tryngites rufescens*).	×	× × ×	×	× ×	×
Great Snipe (*Scolopax major*).	:	× × ×	:	× × × ×	×
White-winged Black Tern (*Sterna leucoptera*).				: ×	:
Whiskered Tern (*Sterna hybrida*).	:	× ×	×	× × ×	×
Gull-billed Tern (*Sterna anglica*).	:	× ×		× × ×	× ×

TABLE VI. SHOWING THE PRINCIPAL OCCURRENCES OF ABNORMAL MIGRANTS.

SPECIES.	JAN.	FEB.	MAR.	APRIL.	MAY.	JUNE.	JULY.	AUG.	SEPT.	OCT.	NOV.	DEC.
Caspian Tern (*Sterna caspia*).	×	×		×	×	×	×	×		×	×	
Sooty Tern (*Sterna fuliginosa*).						×				×		
Smaller Sooty Tern (*Sterna anœstheta*).					×?				×		×	
Noddy Tern (*Sterna stolida*).*				×	×							
Bonaparte's Gull (*Larus philadelphia*).*				×			×					
Mediterranean Black-headed Gull (*Larus melanocephalus*).	×											×
Great Black-headed Gull (*Larus ichthyaetus*).					×?	×?						
Dusky Shearwater (*Puffinus obscurus*).						×						
Sooty Shearwater (*Puffinus griseus*).							×	×			×	
Wilson's Petrel (*Oceanites wilsoni*).					×			×?		×		
Cape Petrel (*Daption capensis*).												
Capped Petrel (*Œstrelata hæsitata*).			×?	×?								
Collared Petrel (*Œstrelata torquata*).												×
Bulwer's Petrel (*Bulweria columbina*).					×							×
Trumpeter Swan (*Cygnus buccinator*).		×										
American Swan (*Cygnus americanus*).								×?		×	×	
Lesser Snow Goose (*Anser hyperboreus*).	×									×	WINTER	
Red-breasted Goose (*Bernicla ruficollis*).	×											
Ruddy Shellrake (*Tadorna rutila*).	×		×				×	×	×	×		
American Wigeon (*Anas americana*).	×								×			
Blue-winged Teal (*Anas discors*).	×		×	×								
American Teal (*Anas carolinensis*).	×	×					×					
Red-crested Pochard (*Fuligula rufina*).	×										WINTER	
White-eyed Pochard (*Fuligula nyroca*).	×	×	SPRING								WINTER	
Buffel-headed Duck (*Clangula albeola*).	×	×									WINTER	
King-necked Duck (*Fuligula collaris*).	×	×										
Hooded Merganser (*Mergus cucullatus*).	×	×										×

NOTE.—Where no signs are affixed the months are unknown, but the season is inserted. An * denotes that the month of capture is unknown.

283

TABLE OF MIGRATION FLIGHT FOR YEAR 189... IN THE AREA OR DISTRICT OF ....

JAN.	FEB.	MAR.	APR.	MAY	JUN.	SPECIES.	JULY	AUG.	SEPT.	OCT.	NOV.	DEC.
						Song Thrush						
						Redwing						
						Fieldfare						
						Ring Ouzel						
						Nightingale						
						Whinchat						
						Redstart						
						Wheatear						
						Spotted Flycatcher						
						Grasshopper Warbler						
						Sedge Warbler						
						Reed Warbler						
						Blackcap						
						Garden Warbler						
						Whitethroat						
						Lesser Whitethroat						
						Wood Wren						
						Willow Wren						
						Chiffchaff						
						Red-backed Shrike						
						Goldcrest						
						Hooded Crow						
						Brambling						
						Snow Bunting						
						Yellow Wagtail						
						Tree Pipit						
						Swallow						
						House Martin						
						Sand Martin						
						Nightjar						
						Cuckoo						
						Swift						
						Wryneck						
						Turtle Dove						
						Corn Crake						
						Stone Curlew						
						Turnstone						
						Golden Plover						
						Whimbrel						
						Common Sandpiper						
						Knot						
						Sanderling						
						Woodcock						
						Jack Snipe						
						Gray-lag Goose						
						Bean Goose						
						Brent Goose						
						Wigeon						
						Tufted Duck						
						Common Scoter						

TABLE OF MIGRATION FLIGHT FOR YEAR 189... IN THE AREA OR DISTRICT OF ....

JAN.	FEB.	MAR.	APR.	MAY	JUN.	SPECIES.	JULY	AUG.	SEPT.	OCT.	NOV.	DEC.
						Song Thrush						
						Redwing						
						Fieldfare						
						Ring Ouzel						
						Nightingale						
						Whinchat						
						Redstart						
						Wheatear						
						Spotted Flycatcher ...						
						Grasshopper Warbler						
						Sedge Warbler						
						Reed Warbler						
						Blackcap						
						Garden Warbler ...						
						Whitethroat						
						Lesser Whitethroat ...						
						Wood Wren						
						Willow Wren						
						Chiffchaff						
						Red-backed Shrike ...						
						Goldcrest						
						Hooded Crow						
						Brambling						
						Snow Bunting						
						Yellow Wagtail ...						
						Tree Pipit						
						Swallow						
						House Martin						
						Sand Martin						
						Nightjar						
						Cuckoo						
						Swift						
						Wryneck						
						Turtle Dove						
						Corn Crake						
						Stone Curlew						
						Turnstone						
						Golden Plover						
						Whimbrel						
						Common Sandpiper ...						
						Knot						
						Sanderling						
						Woodcock						
						Jack Snipe						
						Gray-lag Goose ...						
						Bean Goose						
						Brent Goose						
						Wigeon						
						Tufted Duck						
						Common Scoter ...						

TABLE OF MIGRATION FLIGHT FOR YEAR 189... IN THE AREA OR DISTRICT OF ....

JAN.	FEB.	MAR.	APR.	MAY	JUN.	SPECIES.	JULY	AUG.	SEPT.	OCT.	NOV.	DEC.
						Song Thrush						
						Redwing						
						Fieldfare						
						Ring Ouzel						
						Nightingale						
						Whinchat						
						Redstart						
						Wheatear						
						Spotted Flycatcher ...						
						Grasshopper Warbler						
						Sedge Warbler						
						Reed Warbler						
						Blackcap						
						Garden Warbler ...						
						Whitethroat						
						Lesser Whitethroat ...						
						Wood Wren						
						Willow Wren						
						Chiffchaff						
						Red-backed Shrike ...						
						Goldcrest						
						Hooded Crow						
						Brambling						
						Snow Bunting						
						Yellow Wagtail ...						
						Tree Pipit						
						Swallow						
						House Martin						
						Sand Martin						
						Nightjar						
						Cuckoo						
						Swift						
						Wryneck						
						Turtle Dove						
						Corn Crake						
						Stone Curlew						
						Turnstone						
						Golden Plover						
						Whimbrel						
						Common Sandpiper ...						
						Knot						
						Sanderling						
						Woodcock						
						Jack Snipe						
						Gray-lag Goose ...						
						Bean Goose						
						Brent Goose						
						Wigeon						
						Tufted Duck						
						Common Scoter ...						

TABLE OF MIGRATION FLIGHT FOR YEAR 189... IN THE AREA OR DISTRICT OF ....

JAN.	FEB.	MAR.	APR.	MAY	JUN.	SPECIES.	JULY	AUG.	SEPT.	OCT.	NOV.	DEC.
						Song Thrush						
						Redwing						
						Fieldfare						
						Ring Ouzel						
						Nightingale						
						Whinchat						
						Redstart						
						Wheatear						
						Spotted Flycatcher ...						
						Grasshopper Warbler						
						Sedge Warbler						
						Reed Warbler						
						Blackcap						
						Garden Warbler ...						
						Whitethroat						
						Lesser Whitethroat ...						
						Wood Wren						
						Willow Wren						
						Chiffchaff						
						Red-backed Shrike ...						
						Goldcrest						
						Hooded Crow						
						Brambling						
						Snow Bunting						
						Yellow Wagtail ...						
						Tree Pipit						
						Swallow						
						House Martin						
						Sand Martin						
						Nightjar						
						Cuckoo						
						Swift						
						Wryneck						
						Turtle Dove						
						Corn Crake						
						Stone Curlew						
						Turnstone						
						Golden Plover						
						Whimbrel						
						Common Sandpiper ...						
						Knot						
						Sanderling						
						Woodcock						
						Jack Snipe						
						Gray-lag Goose ...						
						Bean Goose						
						Brent Goose						
						Wigeon						
						Tufted Duck						
						Common Scoter ...						

TABLE OF MIGRATION FLIGHT FOR YEAR 189... IN THE AREA OR DISTRICT OF ....

JAN.	FEB.	MAR.	APR.	MAY	JUN.	SPECIES.	JULY	AUG.	SEPT.	OCT.	NOV.	DEC.
						Song Thrush						
						Redwing						
						Fieldfare						
						Ring Ouzel						
						Nightingale						
						Whinchat						
						Redstart						
						Wheatear						
						Spotted Flycatcher						
						Grasshopper Warbler						
						Sedge Warbler						
						Reed Warbler						
						Blackcap						
						Garden Warbler						
						Whitethroat						
						Lesser Whitethroat						
						Wood Wren						
						Willow Wren						
						Chiffchaff						
						Red-backed Shrike						
						Goldcrest						
						Hooded Crow						
						Brambling						
						Snow Bunting						
						Yellow Wagtail						
						Tree Pipit						
						Swallow						
						House Martin						
						Sand Martin						
						Nightjar						
						Cuckoo						
						Swift						
						Wryneck						
						Turtle Dove						
						Corn Crake						
						Stone Curlew						
						Turnstone						
						Golden Plover						
						Whimbrel						
						Common Sandpiper						
						Knot						
						Sanderling						
						Woodcock						
						Jack Snipe						
						Gray-lag Goose						
						Bean Goose						
						Brent Goose						
						Wigeon						
						Tufted Duck						
						Common Scoter						

TABLE OF MIGRATION FLIGHT FOR YEAR 189... IN THE AREA OR DISTRICT OF ....

JAN.	FEB.	MAR.	APR.	MAY	JUN.	SPECIES.	JULY	AUG.	SEPT.	OCT.	NOV.	DEC.
						Song Thrush						
						Redwing						
						Fieldfare						
						Ring Ouzel						
						Nightingale						
						Whinchat						
						Redstart						
						Wheatear						
						Spotted Flycatcher ...						
						Grasshopper Warbler						
						Sedge Warbler						
						Reed Warbler						
						Blackcap						
						Garden Warbler ...						
						Whitethroat						
						Lesser Whitethroat ...						
						Wood Wren						
						Willow Wren						
						Chiffchaff						
						Red-backed Shrike ...						
						Goldcrest						
						Hooded Crow						
						Brambling						
						Snow Bunting						
						Yellow Wagtail ...						
						Tree Pipit						
						Swallow						
						House Martin						
						Sand Martin						
						Nightjar						
						Cuckoo						
						Swift						
						Wryneck						
						Turtle Dove						
						Corn Crake						
						Stone Curlew						
						Turnstone						
						Golden Plover						
						Whimbrel						
						Common Sandpiper ...						
						Knot						
						Sanderling						
						Woodcock						
						Jack Snipe						
						Gray-lag Goose ...						
						Bean Goose						
						Brent Goose						
						Wigeon						
						Tufted Duck						
						Common Scoter ...						

289

TABLE OF MIGRATION FLIGHT FOR YEAR 189... IN THE AREA OR DISTRICT OF ....

JAN.	FEB.	MAR.	APR.	MAY	JUN.	SPECIES.	JULY	AUG.	SEPT.	OCT.	NOV.	DEC.
						Song Thrush						
						Redwing						
						Fieldfare						
						Ring Ouzel						
						Nightingale						
						Whinchat						
						Redstart						
						Wheatear						
						Spotted Flycatcher ...						
						Grasshopper Warbler						
						Sedge Warbler						
						Reed Warbler						
						Blackcap						
						Garden Warbler ...						
						Whitethroat						
						Lesser Whitethroat ...						
						Wood Wren						
						Willow Wren						
						Chiffchaff						
						Red-backed Shrike ...						
						Goldcrest						
						Hooded Crow						
						Brambling						
						Snow Bunting						
						Yellow Wagtail ...						
						Tree Pipit						
						Swallow						
						House Martin						
						Sand Martin						
						Nightjar						
						Cuckoo						
						Swift						
						Wryneck						
						Turtle Dove						
						Corn Crake						
						Stone Curlew						
						Turnstone						
						Golden Plover						
						Whimbrel						
						Common Sandpiper ...						
						Knot						
						Sanderling						
						Woodcock						
						Jack Snipe						
						Gray-lag Goose ...						
						Bean Goose						
						Brent Goose						
						Wigeon						
						Tufted Duck						
						Common Scoter ...						

U

INDEX.

Abnormal migrants at Heligoland, comparative table of, 184
Abnormal migrants, coincidence of route of, 186
Abnormal migration at Heligoland, 181, 182, 183, 184
Abnormal migration, instances of, in British Islands, 179, 180
Abnormal migration, instances of, in other countries, 181
Accentor alpinus, 140
Accentor modularis, 139
Accentor montanellus, 214
Acredula caudata rosea, 133
Acrocephalus agricola, 184
Acrocephalus palustris, 191
Acrocephalus phragmitis, 21, 119, 151, 208
Acrocephalus turdoides, 56, 262
Ægialitis hiaticula, 23, 211
Ægialitis hiaticula major, 23
Ægialitis jerdoni, 23
Ægialitis minor, 23
Ægialitis nova zelandiæ, 52
Ægialitis semipalmatus, 234
Ægialitis vocifera, 184
Ægialophilus cantianus, 145, 191, 211, 247, 265
Agelæus phœniceus, 184
Alauda arvensis, 155, 230, 234
Alauda cœlivox, 234
Alauda dulcivox, 234
Alauda pispoletta, 184
Alauda sibirica, 232
Alauda tartarica, 184, 232
ALCIDÆ, 49, 165
Algeria, vertical migration in, 136, 137

Ampelis, 166
Anas boschas, 56
Anas crecca, 211
Anas falcata, 106
Anas formosa, 106
ANATIDÆ, 52, 53, 102, 163, 165, 190, 247
Ancient routes, instances of species still following them, 99, 100, 101
Anser, 67
Anser brenta, 216
Anser hyperboreus, 217
Anser segetum, 210, 223
Antarctic breeding-grounds, possible undiscovered, 150
Antarctic regions, geological changes in, 96
Anthus arboreus, 138, 191, 211
Anthus cervinus, 211
Anthus gustavi, 118
Anthus pratensis, 210, 247
Anthus richardi, 232
Arctic America. arrival of migrants in, 216, 217
Arctic birds breeding in the southern hemisphere, 150
Arctic Grouse, nomadic movements of, 162
Arctic regions, autumn in, 228
Ardea cinerea, 181
Argyll, Duke of, on hibernation, etc., 12, 13
Argyll, Duke of, on migration of Song Thrush, 126, 127
Astur atricapillus, 184
Autumn, abundance of birds in, 230

11, *HENRIETTA STREET*, *COVENT GARDEN*, *W.C.*

APRIL, *1892*.

A

Catalogue of Books

PUBLISHED BY

CHAPMAN & HALL

LIMITED.

A separate Illustrated Catalogue is issued, containing

Drawing Examples, Diagrams, Models, Instruments, etc.,

ISSUED UNDER THE AUTHORITY OF

THE SCIENCE AND ART DEPARTMENT,

SOUTH KENSINGTON,

FOR THE USE OF SCHOOLS AND ART AND SCIENCE CLASSES.

NEW AND FORTHCOMING BOOKS.

A MIRROR OF THE TURF; or, The Machinery of Horse-racing Revealed, showing the Sport of Kings as it is to-day By Louis Henry Curzon. Crown 8vo.

FROM SINNER TO SAINT; or, Character Transformations : being a few Biographical Sketches of Historic Individuals whose Moral Lives underwent a Remarkable Change. By John Burn Bailey, Author of "Modern Methuselahs." Crown 8vo.

STUDIES AT LEISURE. By W. L. Courtney, Author of "Studies Old and New," &c. Crown 8vo.

SIBERIA AS IT IS. By H. de Windt, Author of "From Pekin to Calais," "A Ride to India," &c. With an Introduction by Malame Olga Novikoff ("O. K.") With numerous Illustrations. Demy 8vo, 18s.

RUSSIAN CHARACTERISTICS. By E. B. Lanin. Reprinted, with revisions, from *The Fortnightly Review.* Demy 8vo.

THE NATURALIST IN LA PLATA. By W. H. Hudson, C.M.Z.S., Joint Author of "Argentine Ornithology." With numerous Illustrations. Demy 8vo, 16s.

A HISTORY OF ANCIENT ART IN PERSIA. By Georges Perrot and Charles Chipiez. With 254 Engravings and 12 Steel and Coloured Plates. Royal 8vo, 21s.

A HISTORY OF ANCIENT ART IN PHRYGIA—Lydia and Caria—Lycia. By Georges Perrot and Charles Chipiez. With 280 Illustrations. Royal 8vo, 15s.

TRAVELS IN AFRICA DURING THE YEARS 1879 to 1883. By Dr. William Junker. Second Volume. With numerous Full-page Plates and Illustrations in the Text. Translated from the German by Professor Keane. Demy 8vo, 21s.

ON SHIBBOLETHS. By W. S. Lilly. Demy 8vo, 12s.
. *Progress, Liberty, The People, Public Opinion, Education, Women's Rights, and Supply and Demand.*

MY THOUGHTS ON MUSIC AND MUSICIANS. By H. H. Statham. Illustrated with frontispiece of the Entrance-front of Handel's Opera House and Musical Examples. Demy 8vo, 18s.

LIFE IN ANCIENT EGYPT AND ASSYRIA. By G. Maspéro, late Director of Archæology in Egypt and Member of the Institute of France. Translated by A. P. Morton. With 188 Illustrations. Crown 8vo, 5s.

AMONG TYPHOONS AND PIRATE CRAFT. By Captain Lindsay Anderson, Author of "A Cruise in an Opium Clipper." With Illustrations. Crown 8vo, 5s.

AUSTRALIAN LIFE. By Francis Adams. Crown 8vo, 3s. 6d.

ELINE VERE. By Louis Couperus. Translated by J. T. Grein. Crown 8vo, 5s.

A PARTNER FROM THE WEST. By Arthur Paterson. Crown 8vo, 5s.

PRETTY MICHAL. By Maurice Jokäi. Translated by R. Nisbet Bain. Crown 8vo, 5s.

BOOKS

PUBLISHED BY

CHAPMAN & HALL, LIMITED.

ABOUT (EDMOND)—
HANDBOOK OF SOCIAL ECONOMY; OR, THE
WORKER'S A B C. From the French. With a Biographical and Critical
Introduction by W. FRASER RAE. Second Edition, revised. Crown 8vo, 4s.

ADAMS (FRANCIS)—
AUSTRALIAN LIFE. Crown 8vo, 3s. 6d.

AFRICAN FARM, STORY OF AN. By OLIVE SCHREINER
(Ralph Iron). New Edition. Crown 8vo, 1s.; in cloth, 1s. 6d.
A NEW EDITION, on Superior Paper, and Strongly Bound in Cloth, 3s. 6d.

AGRICULTURAL SCIENCE (LECTURES ON), AND
OTHER PROCEEDINGS OF THE INSTITUTE OF AGRICULTURE,
SOUTH KENSINGTON, 1883-4. Crown 8vo, sewed, 2s.

ANDERSON (ANDREW A.)—
A ROMANCE OF N'SHABÉ: Being a Record of Startling
Adventures in South Central Africa. With Illustrations. Crown 8vo, 5s.

TWENTY-FIVE YEARS IN A WAGGON IN THE
GOLD REGIONS OF AFRICA. With Illustrations and Map. Second Edition.
Demy 8vo, 12s.

ANDERSON (CAPTAIN LINDSAY)—
AMONG TYPHOONS AND PIRATE CRAFT.
With Illustrations by STANLEY WOOD. Crown 8vo, 5s.

A CRUISE IN AN OPIUM CLIPPER. With Illustrations. Crown 8vo, 6s.

ASTRONOMY (A NEW DEPARTURE IN). THE RE-
VOLUTION OF THE SOLAR SYSTEM. By E. H. Demy 8vo, sewed, 2s.

AVELING (EDWARD), D.Sc., Fellow of University College, London—
MECHANICS AND EXPERIMENTAL SCIENCE.
As required for the Matriculation Examination of the University of London.
MECHANICS. With numerous Woodcuts. Crown 8vo, 6s.
Key to Problems in ditto, crown 8vo, 3s. 6d.
CHEMISTRY. With numerous Woodcuts. Crown 8vo, 6s.
Key to Problems in ditto, crown 8vo, 2s. 6d.
MAGNETISM AND ELECTRICITY. With Numerous Woodcuts.
Crown 8vo. 6s.
LIGHT AND HEAT. With Numerous Woodcuts. Crown 8vo, 6s.
Keys to the last two volumes in one vol. Crown 8vo, 5s.

BAILEY (JOHN BURN)—
FROM SINNER TO SAINT; OR, CHARACTER TRANS-
FORMATIONS: being a few Biographical Sketches of Historic Individuals whose
Moral Lives underwent a Remarkable Change. Crown 8vo.

MODERN METHUSELAHS; or, Short Biographical
Sketches of a few advanced Nonagenarians or actual Centenarians. Demy 8vo,
10s. 6d.

BEATTY-KINGSTON (W.)—
A JOURNALIST'S JOTTINGS. 2 vols. Demy 8vo, 24s.

MY "HANSOM" LAYS: Original Verses, Imitations, and
Paraphrases. Crown 8vo, 3s. 6d.

THE CHUMPLEBUNNYS AND SOME OTHER
ODDITIES. Sketched from the Life. Illustrated by KARL KLIETSCH. Crown
8vo, 2s.; paper, 1s.

A WANDERER'S NOTES. 2 vols. Demy 8vo, 24s.

MONARCHS I HAVE MET. 2 vols. Demy 8vo, 24s.

MUSIC AND MANNERS: Personal Reminiscences and
Sketches of Character. 2 vols. Demy 8vo, 30s.

A 2

BELL (JAMES, Ph.D., &c.), Principal of the Somerset House Laboratory—
THE CHEMISTRY OF FOODS.   With Microscopic
Illustrations.
PART I. TEA, COFFEE, COCOA, SUGAR, ETC.   Large crown 8vo, 2s. 6d.
PART II.   MILK, BUTTER, CHEESE, CEREALS, PREPARED
STARCHES, ETC. Large crown 8vo, 3s.

BENSON (W.)—
UNIVERSAL PHONOGRAPHY. To classify sounds of
Human Speech, and to denote them by one set of Symbols for easy Writing and
Printing. 8vo, sewed, 1s.
MANUAL OF THE SCIENCE OF COLOUR. Coloured
Frontispiece and Illustrations. 12mo; cloth, 2s. 6d.
PRINCIPLES OF THE SCIENCE OF COLOUR. Small
4to, cloth, 15s.

BIRDWOOD (SIR GEORGE C. M.), C.S.I.—
THE INDUSTRIAL ARTS OF INDIA. With Map and
174 Illustrations. New Edition. Demy 8vo, 14s.

BLACKIE (JOHN STUART), F.R.S.E.—
THE SCOTTISH HIGHLANDERS AND THE LAND
LAWS. Demy 8vo, 9s.
ALTAVONA: FACT AND FICTION FROM MY LIFE
IN THE HIGHLANDS. Third Edition. Crown 8vo, 6s.

BLEUNARD (A.)—
BABYLON ELECTRIFIED : The History of an Expe-
dition undertaken to restore Ancient Babylon by the Power of Electricity, and how
it Resulted. Translated from the French by F. L. WHITE, and Illustrated by
MONTADER. Royal 8vo, 21s.

BLOOMFIELD'S (BENJAMIN LORD), MEMOIR OF—
MISSION TO THE COURT OF BERNADOTTE. With Portraits. 2 vols.
Demy 8vo, 28s.

BONVALOT (GABRIEL)—
THROUGH THE HEART OF ASIA OVER THE
PAMIR TO INDIA. Translated from the French by C. B. PITMAN. With
250 Illustrations by ALBERT PÉPIN. Royal 8vo, 32s.

BOWERS (G.)—
HUNTING IN HARD TIMES. With 61 coloured
Illustrations. Oblong 4to, 12s.

BRACKENBURY (COL. C. B.)—
FREDERICK THE GREAT. With Maps and Portrait.
Large crown 8vo, 4s.

BRADLEY (THOMAS), of the Royal Military Academy, Woolwich—
ELEMENTS OF GEOMETRICAL DRAWING. In Two
Parts, with Sixty Plates. Oblong folio, half bound, each Part 16s.

BRIDGMAN (F. A.)—
WINTERS IN ALGERIA. With 62 Illustrations. Royal
8vo, 10s. 6d.

BRITISH ARMY, THE. By the Author of "Greater Britain,"
" The Present Position of European Politics," etc. Demy 8vo, 12s.

BROCK (DR. J. H. E.), Assistant Examiner in Hygiene, Science and Art Department—
ELEMENTS OF HUMAN PHYSIOLOGY FOR THE
HYGIENE EXAMINATIONS OF THE SCIENCE AND ART
DEPARTMENT. Crown 8vo, 1s. 6d.

BROMLEY-DAVENPORT (the late W.), M.P.—
SPORT: Fox Hunting, Salmon Fishing, Covert Shooting,
Deer Stalking. With numerous Illustrations by General CREALOCK, C.B.
New Cheap Edition. Post 8vo, 3s. 6d.

BROWN (J. MORAY)—
POWDER, SPEAR, AND SPUR: A Sporting Medley.
With Illustrations. Crown 8vo, 10s. 6d.

BUCKLAND (FRANK)—
LOG-BOOK OF A FISHERMAN AND ZOOLOGIST.
With numerous Illustrations. Sixth Thousand. Crown 8vo, 3s. 6d.

BUCKMAN (S. S.), F.G.S.—
ARCADIAN LIFE. With Illustrations by P. BUCKMAN.
Crown 8vo., 1s.
JOHN DARKE'S SOJOURN IN THE COTTESWOLDS
AND ELSEWHERE : A Series of Sketches. Crown 8vo, 1s.

BURCHETT (R.)—
DEFINITIONS OF GEOMETRY. New Edition. 24mo,
cloth, 5d.
LINEAR PERSPECTIVE, for the Use of Schools of Art.
New Edition. With Illustrations. Post 8vo, cloth, 7s.
PRACTICAL GEOMETRY: The Course of Construction
of Plane Geometrical Figures. With 137 Diagrams. Eighteenth Edition. Post
8vo, cloth, 5s.

BURGESS (EDWARD)—
ENGLISH AND AMERICAN YACHTS. Illustrated
with 50 Beautiful Photogravure Engravings. Oblong folio, 42s.

BUTLER (A. J.)—
COURT LIFE IN EGYPT. Second Edition. Illustrated.
Large crown 8vo, 12s.

CAFFYN (MANNINGTON)—
A POPPY'S TEARS. Crown 8vo, 1s. ; in cloth, 1s. 6d.

CARLYLE (THOMAS), WORKS BY.—See pages 29 and 30.
THE CARLYLE BIRTHDAY BOOK. Compiled, with
the permission of Mr. Thomas Carlyle, by C. N. WILLIAMSON. Second Edition.
Small fcap. 8vo, 3s.

CARSTENSEN (A. RIIS)—
TWO SUMMERS IN GREENLAND: An Artist's
Adventures among Ice and Islands in Fjords and Mountains. With numerous
Illustrations by the Author. Demy 8vo, 14s.

CHAPMAN & HALL'S SHILLING SERIES.

THE CHUMPLEBUNNYS AND SOME OTHER ODDITIES. Sketched from Life. By W. BEATTY-KINGSTON. Illustrated by KARL KLIETSCH. Crown 8vo.

A SUBURB OF YEDO. By the late THEOBALD A. PURCELL. With Numerous Illustrations.

ARCADIAN LIFE. By S. S. BUCKMAN, F.G.S. With Illustrations. Crown 8vo.

* JOHN DARKE'S SOJOURN IN THE COTTESWOLDS AND ELSEWHERE. By S. S. BUCKMAN. With Illustrations.

SINGER'S STORY, A. Related by the Author of "Flitters, Tatters, and the Counsellor."

* A POPPY'S TEARS By MANNINGTON CAFFYN.

* NOTCHES ON THE ROUGH EDGE OF LIFE. By LYNN CYRIL D'OYLE.

* WE TWO AT MONTE CARLO. By ALBERT D. VANDAM.

* WHO IS THE MAN? A Tale of the Scottish Border. By J S. TAIT.

* THE CHILD OF STAFFERTON. By CANON KNOX LITTLE.

THE BROKEN VOW : A Story of Here and Hereafter. By CANON KNOX LITTLE.

* THE STORY OF AN AFRICAN FARM. By OLIVE SCHREINER.

* PADDY AT HOME. By BARON E. DE MANDAT-GRANCEY.

* Bound in Cloth, 1s. 6d.

CHARLOTTE ELIZABETH, LIFE AND LETTERS OF,
Princess Palatine and Mother of Philippe d'Orléans, Regent of France, 1652–1722. With Portraits. Demy 8vo, 10s. 6d.

CHARNAY (DÉSIRÉ)—

THE ANCIENT CITIES OF THE NEW WORLD.
Being Travels and Explorations in Mexico and Central America, 1857—1882. With upwards of 200 Illustrations. Super Royal 8vo, 31s. 6d.

CHRIST THAT IS TO BE, THE : A Latter-day Romance.
Third Edition. Demy 8vo, 3s. 6d.

CHURCH (PROFESSOR A. H.), M.A. Oxon.—

FOOD GRAINS OF INDIA. With numerous Woodcuts.
Small 4to, 6s.

ENGLISH PORCELAIN. A Handbook to the China
made in England during the Eighteenth Century, as illustrated by Specimens chiefly in the National Collection. With numerous Woodcuts. Large crown 8vo, 3s.

ENGLISH EARTHENWARE. A Handbook to the
Wares made in England during the 17th and 18th Centuries, as illustrated by Specimens in the National Collections. With numerous Woodcuts. Large crown 8vo, 3s.

PLAIN WORDS ABOUT WATER. Illustrated. Crown
8vo, sewed, 6d.

FOOD : Some Account of its Sources, Constituents, and
Uses. A New and Revised Edition. Large crown 8vo, cloth, 3s.

PRECIOUS STONES : considered in their Scientific and
Artistic Relations. With a Coloured Plate and Woodcuts. Second Edition. Large crown 8vo, 2s. 6d.

COBDEN, RICHARD, LIFE OF. By the Right Hon. John
Morley, M.P. With Portrait. New Edition. Crown 8vo, 7s. 6d.
Popular Edition, with Portrait, 4to, sewed, 1s.; cloth, 2s.

COLLINS (WILKIE) and DICKENS (CHARLES)—
THE LAZY TOUR OF TWO IDLE APPRENTICES;
NO THOROUGHFARE; THE PERILS OF CERTAIN ENGLISH
PRISONERS. With 8 Illustrations. Crown 8vo, 5s.
. These Stories are now reprinted for the first time complete.

COOKERY—
DINNERS IN MINIATURE. By Mrs. Earl. Crown
8vo, 2s. 6d.

HILDA'S "WHERE IS IT?" OF RECIPES. Contain-
ing many old CAPE, INDIAN, and MALAY DISHES and PRESERVES;
also Directions for Polishing Furniture, Cleaning Silk, etc.; and a Collection of
Home Remedies in Case of Sickness. By Hildagonda J. Duckitt. Inter-
leaved with White Paper for adding Recipes. Third Edition. Crown 8vo, 4s. 6d.

THE PYTCHLEY BOOK OF REFINED COOKERY
AND BILLS OF FARE. By Major L——. Fourth Edition. Large crown 8vo,
8s.

BREAKFASTS, LUNCHEONS, AND BALL SUPPERS.
By Major L——. Crown 8vo. 4s.

OFFICIAL HANDBOOK OF THE NATIONAL
TRAINING SCHOOL FOR COOKERY. Containing Lessons on Cookery
forming the Course of Instruction in the School. Compiled by "R. O. C."
Twenty-first Thousand. Large crown 8vo, 6s.

BREAKFAST AND SAVOURY DISHES. By "R. O. C."
Ninth Thousand. Crown 8vo 1s.

HOW TO COOK FISH. Compiled by "R. O. C."
Crown 8vo, sewed, 3d.

SICK-ROOM COOKERY. Compiled by "R. O. C."
Crown 8vo, sewed, 6d.

THE ROYAL CONFECTIONER: English and Foreign.
A Practical Treatise. By C. E. Francatelli. With numerous Illustrations.
Sixth Thousand. Crown 8vo, 5s.

THE KINGSWOOD COOKERY BOOK. By H. F.
WICKEN. Crown 8vo, 2s.

COOPER-KING (LT.-COL.)—
GEORGE WASHINGTON. Large crown 8vo. With
Portrait and Maps. [*In the Press.*

COUPERUS (LOUIS)—
ELINE VERE. Translated from the Dutch by J. T.
Grein. Crown 8vo, 5s.

COURTNEY (W. L.), M.A., LL.D., of New College, Oxford—
STUDIES AT LEISURE. Crown 8vo. [*In the Press.*

STUDIES NEW AND OLD. Crown 8vo, 6s.

CONSTRUCTIVE ETHICS: A Review of Modern Philo-
sophy and its Three Stages of Interpretation, Criticism, and Reconstruction.
Demy 8vo, 12s.

CRAIK (GEORGE LILLIE)—
ENGLISH OF SHAKESPEARE. Illustrated in a Philological Commentary on "Julius Cæsar." Eighth Edition. Post 8vo, cloth, 5s.
OUTLINES OF THE HISTORY OF THE ENGLISH LANGUAGE. Eleventh Edition. Post 8vo, cloth, 2s. 6d.

CRAWFURD (OSWALD)—
ROUND THE CALENDAR IN PORTUGAL. With numerous Illustrations. Royal 8vo, 18s.
BEYOND THE SEAS; being the surprising Adventures and ingenious Opinions of Ralph, Lord St. Keyne, told by his kinsman, Humphrey St. Keyne. Second Edition. Crown 8vo, 3s. 6d.

CRIPPS (WILFRED JOSEPH), M.A., F.S.A.—
COLLEGE AND CORPORATION PLATE. A Handbook for the Reproduction of Silver Plate. With numerous Illustrations. Large crown 8vo, cloth, 2s. 6d.

CURZON (LOUIS HENRY)—
A MIRROR OF THE TURF; or, The Machinery of Horse-racing Revealed, showing the Sport of Kings as it is to-day. Crown 8vo.

DAIRY FARMING—
DAIRY FARMING. To which is added a Description of the Chief Continental Systems. With numerous Illustrations. By JAMES LONG. Crown 8vo, 9s.
DAIRY FARMING, MANAGEMENT OF COWS, etc. By ARTHUR ROLAND. Edited by WILLIAM ABLETT. Crown 8vo, 5s.

DALY (J. B.), LL.D.—
IRELAND IN THE DAYS OF DEAN SWIFT. Crown 8vo, 5s.

DAS (DEVENDRA N.)—
SKETCHES OF HINDOO LIFE. Crown 8vo, 5s.

DAUBOURG (E.)—
INTERIOR ARCHITECTURE. Doors, Vestibules, Staircases, Anterooms, Drawing, Dining, and Bed Rooms, Libraries, Bank and Newspaper Offices, Shop Fronts and Interiors. Half-imperial, cloth, £2 12s. 6d.

DAVIDSON (ELLIS A.)—
PRETTY ARTS FOR THE EMPLOYMENT OF LEISURE HOURS. A Book for Ladies. With Illustrations. Demy 8vo, 6s.

DAY (WILLIAM)—
THE RACEHORSE IN TRAINING, with Hints on Racing and Racing Reform, to which is added a Chapter on Shoeing. Sixth Edition. Demy 8vo, 9s.

DE BOVET (MADAME)—
THREE MONTHS' TOUR IN IRELAND. Translated and Condensed by MRS. ARTHUR WALTER. With Illustrations. Crown 8vo., 6s.

DE CHAMPEAUX (ALFRED)—
TAPESTRY. With numerous Woodcuts. Cloth, 2s. 6d.

DE FALLOUX (THE COUNT)—
MEMOIRS OF A ROYALIST. Edited by C. B. PITMAN. 2 vols. With Portraits. Demy 8vo, 32s.

DE KONINCK (L. L.) and DIETZ (E.)—
PRACTICAL MANUAL OF CHEMICAL ASSAYING, as applied to the Manufacture of Iron. Edited, with notes, by ROBERT MALLET. Post 8vo, cloth, 6s.

DE LESSEPS (FERDINAND)—
RECOLLECTIONS OF FORTY YEARS. Translated
from the French by C. B. PITMAN. 2 vols. Demy 8vo, 24s.

DE LISLE (MEMOIR OF LIEUTENANT RUDOLPH),
R.N., of the Naval Brigade. By the Rev. H. N. OXENHAM, M.A. Third
Edition. Crown 8vo, 7s. 6d.

DE MANDAT-GRANCEY (BARON E.)—
PADDY AT HOME; OR, IRELAND AND THE IRISH AT
THE PRESENT TIME, AS SEEN BY A FRENCHMAN. Fifth Edition. Crown 8vo, 1s. ;
in cloth, 1s. 6d.

D'OYLE (LYNN CYRIL)—
NOTCHES ON THE ROUGH EDGE OF LIFE.
Crown 8vo, 1s. ; in cloth, 1s. 6d.

DE STAËL (MADAME)—
MADAME DE STAËL: Her Friends, and Her Influence
in Politics and Literature. By LADY BLENNERHASSETT. Translated from the
German by J. E. GORDON CUMMING. With a Portrait. 3 vols. Demy 8vo, 36s.

DE WINDT (H.)—
SIBERIA AS IT IS. With an Introduction by MADAME
OLGA NOVIKOFF ("O. K.") With numerous Illustrations. Demy 8vo, 18s.

FROM PEKIN TO CALAIS BY LAND. With nume-
rous Illustrations by C. E. FRIPP from Sketches by the Author. Demy 8vo, 20s.

A RIDE TO INDIA ACROSS PERSIA AND BELU-
CHISTAN. With numerous Illustrations and Map. Demy 8vo, 16s.

DICKENS (CHARLES), WORKS BY—See pages 31—37.
THE LETTERS OF CHARLES DICKENS. Two
vols. uniform with "The Charles Dickens Edition" of his Works. Crown 8vo, 7s.

THE LIFE OF CHARLES DICKENS—*See "Forster."*

THE CHARLES DICKENS BIRTHDAY BOOK.
With Five Illustrations. In a handsome fcap. 4to volume, 12s.

THE HUMOUR AND PATHOS OF CHARLES
DICKENS. By CHARLES KENT. With Portrait. Crown 8vo, 6s.

THE DICKENS DICTIONARY. A Key to the Charac-
ters and Principal Incidents in the Tales of Charles Dickens. By GILBERT
PIERCE, with Additions by WILLIAM A. WHEELER. New Edition, uniform with
the "Crown" Edition of Dickens's Works. Large crown 8vo, 5s.

DICKENS (CHARLES) and COLLINS (WILKIE)—
THE LAZY TOUR OF TWO IDLE APPRENTICES;
NO THOROUGHFARE; THE PERILS OF CERTAIN ENGLISH
PRISONERS. With Illustrations. Crown 8vo, 5s.
⁎⁎ These Stories are now reprinted in complete form for the first time.

DILKE (LADY)—
ART IN THE MODERN STATE. With Facsimile.
Demy 8vo, 9s.

DINARTE (SYLVIO)—
INNOCENCIA : A Story of the Prairie Regions of Brazil.
Translated from the Portuguese and Illustrated by JAMES W. WELLS, F.R.G.S.
Crown 8vo, 6s.

DIXON (CHARLES)—
THE BIRDS OF OUR RAMBLES : A Companion
for the Country. With Illustrations by A. T. ELWES. Large Crown 8vo, 7s. 6d.

IDLE HOURS WITH NATURE. With Frontispiece.
Crown 8vo, 6s.

ANNALS OF BIRD LIFE : A Year-Book of British
Ornithology. With Illustrations. Crown 8vo, 7s. 6d.

B

10 *BOOKS PUBLISHED BY*

DOUGLAS (JOHN)—
SKETCH OF THE FIRST PRINCIPLES OF PHYSIO-
GRAPHY. With Maps and numerous Illustrations. Crown 8vo, 6s.

DRAYSON (MAJOR-GENERAL A. W.)—
THIRTY THOUSAND YEARS OF THE EARTH'S
PAST HISTORY. Large Crown 8vo, 5s.
EXPERIENCES OF A WOOLWICH PROFESSOR
during Fifteen Years at the Royal Military Academy. Demy 8vo, 8s.
PRACTICAL MILITARY SURVEYING AND
SKETCHING. Fifth Edition. Post 8vo, cloth, 4s. 6d.

DUCKITT (HILDAGONDA J.)—
HILDA'S "WHERE IS IT?" OF RECIPES. Contain-
ing many old CAPE, INDIAN, and MALAY DISHES and PRESERVES;
also Directions for Polishing Furniture, Cleaning Silk, etc. Third Edition.
Crown 8vo, 4s. 6d.

DUCOUDRAY (GUSTAVE)—
THE HISTORY OF ANCIENT CIVILISATION. A
Handbook based upon M. Gustave Ducoudray's "Histoire Sommaire de la
Civilisation." Edited by REV. J. VERSCHOYLE, M.A. With Illustrations. Large
crown 8vo, 6s.
THE HISTORY OF MODERN CIVILISATION. With
Illustrations. Large crown 8vo, 9s.

DUFFY (SIR CHARLES GAVAN), K.C.M.G.—
THE LEAGUE OF NORTH AND SOUTH. An Episode
in Irish History, 1850–1854. Crown 8vo, 8s.

DYCE (WILLIAM), R.A.—
DRAWING-BOOK OF THE GOVERNMENT SCHOOL
OF DESIGN. Fifty selected Plates. Folio, sewed, 5s.; mounted, 18s.
ELEMENTARY OUTLINES OF ORNAMENT. Plates I.
to XXII., containing 97 Examples, adapted for Practice of Standards I. to IV.
Small folio, sewed, 2s 6d.
SELECTION FROM DYCE'S DRAWING BOOK.
15 Plates, sewed, 1s. 6d.; mounted on cardboard, 6s. 6d.
TEXT TO ABOVE. Crown 8vo, sewed, 6d.
DYNAMIC ACTION AND PONDEROSITY OF MATTER
(FRESH LIGHT ON THE). By WATERDALE. Crown 8vo, 2s. 6d.

EARL. (MRS.)—
DINNERS IN MINIATURE. Crown 8vo, 2s. 6d.

EDWARDS (MRS. SUTHERLAND)—
THE SECRET OF THE PRINCESS. A Tale of
Country, Camp, Court, Convict, and Cloister Life in Russia. Crown 8vo, 3s. 6d.

ELLIS (A. B., Major 1st West India Regiment)—
THE EWE-SPEAKING PEOPLE OF THE SLAVE
COAST OF WEST AFRICA. With Map. Demy 8vo, 10s. 6d.
THE TSHI-SPEAKING PEOPLES OF THE GOLD
COAST OF WEST AFRICA: their Religion, Manners, Customs, Laws,
Language, &c. With Map. Demy 8vo, 10s. 6d.
SOUTH AFRICAN SKETCHES. Crown 8vo, 6s.
THE HISTORY OF THE WEST INDIA REGI-
MENT. With Maps and Coloured Frontispiece and Title-page. Demy 8vo, 18s.
THE LAND OF FETISH. Demy 8vo, 12s.

ENGEL (CARL)—
MUSICAL INSTRUMENTS. With numerous Woodcuts.
Large crown 8vo, cloth, 2s. 6d.

ESCOTT (T. H. S.)—
POLITICS AND LETTERS. Demy 8vo, 9s.

ENGLAND: ITS PEOPLE, POLITY, AND PURSUITS.
New and Revised Edition. Demy 8vo, 3s. 6d.

EUROPEAN POLITICS, THE PRESENT POSITION OF.
By the Author of "Greater Britain." Demy 8vo, 12s.

FANE (VIOLET)—
AUTUMN SONGS. Crown 8vo, 6s.

THE STORY OF HELEN DAVENANT. Crown 8vo,
3s. 6d.

QUEEN OF THE FAIRIES (A Village Story), and other
Poems. Crown 8vo, 6s.

ANTHONY BABINGTON: a Drama. Crown 8vo, 6s.

FARR (WILLIAM) and THRUPP (GEORGE A.)—
COACH TRIMMING. With 60 Illustrations. Crown 8vo,
2s. 6d.

FIELD (HENRY M.)—
GIBRALTAR. With numerous Illustrations. Demy 8vo,
7s. 6d.

FITZGERALD (PERCY), F.S.A.—
THE HISTORY OF PICKWICK. An Account of its
Characters, Localities, Allusions, and Illustrations. With a Bibliography. Demy
8vo, 8s.
A few copies are issued with impressions from the First Set of Steel Plates, 14s.

FLEMING (GEORGE), F.R.C.S.—
ANIMAL PLAGUES: THEIR HISTORY, NATURE,
AND PREVENTION. 8vo, cloth, 15s.

PRACTICAL HORSE-SHOEING. With 37 Illustrations.
Fifth Edition, enlarged. 8vo, sewed, 2s.

RABIES AND HYDROPHOBIA: THEIR HISTORY,
NATURE, CAUSES, SYMPTOMS, AND PREVENTION. With 8 Illustrations. 8vo, cloth, 15s.

FORSTER (JOHN)—
THE LIFE OF CHARLES DICKENS. Original
Edition. Vol. I., 8vo, cloth, 12s. Vol. II., 8vo, cloth, 14s. Vol. III., 8vo, cloth,
16s.
Uniform with the Illustrated Library Edition of Dickens's
Works. 2 vols. Demy 8vo, 20s.
Uniform with the Library and Popular Library Editions.
Post 8vo, 10s. 6d. each.
Uniform with the "C. D." Edition. With Numerous
Illustrations. 2 vols. 7s.
Uniform with the Crown Edition. Crown 8vo. [In the Press.
Uniform with the Household Edition. With Illustrations
by F. BARNARD. Crown 4to, cloth, 5s.

FORSTER, THE LIFE OF THE RIGHT HON. W. E.
By T. WEMYSS REID. With Portraits. Fourth Edition. 2 vols. Demy 8vo, 32s.
FIFTH EDITION in one volume with new Portrait. Demy 8vo, 6d.

B 2

FORSYTH (CAPTAIN)—
THE HIGHLANDS OF CENTRAL INDIA: Notes on
their Forests and Wild Tribes, Natural History and Sports. With Map and
Coloured Illustrations. A New Edition. Demy 8vo, 12s.

FORTNIGHTLY REVIEW (see page 40)—
FORTNIGHTLY REVIEW.—First Series, May, 1865, to
Dec. 1866. 6 vols. Cloth, 13s. each.
New Series, 1867 to 1872. In Half-yearly Volumes. Cloth,
13s. each.
From January, 1873, to the present time, in Half-yearly
Volumes. Cloth, 16s. each.
CONTENTS OF FORTNIGHTLY REVIEW. From
the commencement to end of 1878. Sewed, 2s.

FORTNUM (C. D. E.), F.S.A.—
MAIOLICA. With numerous Woodcuts. Large crown
8vo, cloth, 2s. 6d.
BRONZES. With numerous Woodcuts. Large crown
8vo, cloth, 2s. 6d.

FOUQUÉ (DE LA MOTTE)—
UNDINE: a Romance translated from the German. With
an Introduction by JULIA CARTWRIGHT. Illustrated by HEYWOOD SUMNER.
Crown 4to, 5s.

FRANCATELLI (C. E.)—
THE ROYAL CONFECTIONER: English and Foreign.
A Practical Treatise. With Illustrations. Sixth Thousand. Crown 8vo, 5s.

FRANKS (A. W.)—
JAPANESE POTTERY. Being a Native Report, with an
Introduction and Catalogue. With numerous Illustrations and Marks. Large
crown 8vo, cloth, 2s. 6d.

FROBEL, FRIEDRICH; a Short Sketch of his Life, including
Fröbel's Letters from Dresden and Leipzig to his Wife, now first Translated into
English. By EMILY SHIRREFF. Crown 8vo, 2s.

GALLENGA (ANTONIO)—
ITALY: PRESENT AND FUTURE. 2 vols. Dmy. 8vo, 21s.
EPISODES OF MY SECOND LIFE. 2 vols. Dmy. 8vo, 28s.
IBERIAN REMINISCENCES. Fifteen Years' Travelling
Impressions of Spain and Portugal. With a Map. 2 vols. Demy 8vo, 32s.

GASNAULT (PAUL) and GARNIER (ED.)—
FRENCH POTTERY. With Illustrations and Marks.
Large crown 8vo, 3s.

GILLMORE (PARKER)—
THE HUNTER'S ARCADIA. With numerous Illustrations. Demy 8vo, 10s. 6d.

GIRL'S LIFE EIGHTY YEARS AGO (A). Selections from
the Letters of Eliza Southgate Bowne, with an Introduction by Clarence Cook.
Illustrated with Portraits and Views. Crown 4to, 12s.

GLEICHEN (COUNT), Grenadier Guards—
WITH THE CAMEL CORPS UP THE NILE. With
numerous Sketches by the Author. Third Edition. Large crown 8vo, 9s.

GORDON (GENERAL)—
LETTERS FROM THE CRIMEA, THE DANUBE,
AND ARMENIA. Edited by DEMETRIUS C. BOULGER. Second Edition.
Crown 8vo, 5s.

GORST (SIR J. E.), Q.C., M.P.—
An ELECTION MANUAL. Containing the Parliamentary
Elections (Corrupt and Illegal Practices) Act, 1883, with Notes. Third Edition.
Crown 8vo, 1s. 6d.

GOWER (A. R.), Royal School of Mines—
PRACTICAL METALLURGY. With Illustrations. Crown
8vo, 3s.

GRESWELL (WILLIAM), M.A., F.R.C.I.—
OUR SOUTH AFRICAN EMPIRE. With Map. 2 vols.
Crown 8vo, 21s.

GRIFFIN (SIR LEPEL HENRY), K.C.S.I.—
THE GREAT REPUBLIC. Second Edition. Crown 8vo,
4s. 6d.

GRIFFITHS (MAJOR ARTHUR), H.M. Inspector of Prisons—
FRENCH REVOLUTIONARY GENERALS. Large
crown 8vo, 6s.
CHRONICLES OF NEWGATE. Illustrated. New
Edition. Demy 8vo, 16s.
MEMORIALS OF MILLBANK: or, Chapters in Prison
History. With Illustrations. New Edition. Demy 8vo, 12s.

HALL (SIDNEY)—
A TRAVELLING ATLAS OF THE ENGLISH COUN-
TIES. Fifty Maps, coloured. New Edition, including the Railways, corrected
up to the present date. Demy 8vo, in roan tuck, 10s. 6d.

HAWKINS (FREDERICK)—
THE FRENCH STAGE IN THE EIGHTEENTH
CENTURY. With Portraits. 2 vols. Demy 8vo, 30s.
ANNALS OF THE FRENCH STAGE: FROM ITS
ORIGIN TO THE DEATH OF RACINE. 4 Portraits. 2 vols. Demy 8vo, 28s.

HILDEBRAND (HANS), Royal Antiquary of Sweden—
INDUSTRIAL ARTS OF SCANDINAVIA IN THE
PAGAN TIME. With numerous Woodcuts. Large crown 8vo, 2s. 6d.

HILL (MISS G.)—
THE PLEASURES AND PROFITS OF OUR LITTLE
POULTRY FARM. Small 8vo, 3s.

HOLBEIN—
TWELVE HEADS AFTER HOLBEIN. Selected from
Drawings in Her Majesty's Collection at Windsor. Reproduced in Autotype, in
portfolio. £1 16s.

HOLMES (GEORGE C. V.), Secretary of the Institution of Naval Architects—
MARINE ENGINES AND BOILERS. With Sixty-nine
Woodcuts. Large crown 8vo, 3s.
HOPE (ANDRÉE)—
CHRONICLES OF AN OLD INN; or, a Few Words
about Gray's Inn. Crown 8vo, 5s.
HOUSSAYE (ARSÈNE)—
BEHIND THE SCENES OF THE COMÉDIE FRAN-
CAISE, AND OTHER RECOLLECTIONS. Translated from the French.
Demy 8vo, 14s.
HOVELACQUE (ABEL)—
THE SCIENCE OF LANGUAGE: LINGUISTICS,
PHILOLOGY, AND ETYMOLOGY. With Maps. Large crown 8vo, cloth, 5s.
HOZIER (H. M.)—
TURENNE. With Portrait and Two Maps. Large crown
8vo, 4s.
HUDSON (W. H.), C.M.Z.S. Joint Author of " Argentine Ornithology"—
THE NATURALIST IN LA PLATA. With numerous
Illustrations. Demy 8vo, 16s.
HUEFFER (F.)—
HALF A CENTURY OF MUSIC IN ENGLAND.
1837—1887. Demy 8vo, 8s.
HUGHES (W. R.), F.L.S.—
A WEEK'S TRAMP IN DICKENS-LAND. With
upwards of 100 Illustrations by F. G. KITTON, HERBERT RAILTON, and others.
Demy 8vo, 16s.
HUNTLY (MARQUIS OF)—
TRAVELS, SPORTS, AND POLITICS IN THE EAST
OF EUROPE. With Illustrations by the Marchioness of Huntly. Large
Crown 8vo, 12s.
INDUSTRIAL ARTS: Historical Sketches. With numerous
Illustrations. Large crown 8vo, 3s.
JACKSON (FRANK G.), Master in the Birmingham Municipal School of Art—
DECORATIVE DESIGN. An Elementary Text Book of
Principles and Practice. With numerous Illustrations. Second Edition. Crown
8vo, 7s. 6d.
JAMES (HENRY A.), M.A.—
HANDBOOK TO PERSPECTIVE. Crown 8vo, 2s. 6d.
PERSPECTIVE CHARTS, for use in Class Teaching. 2s.
JARRY (GENERAL)—
OUTPOST DUTY. With TREATISES ON MILITARY
RECONNAISSANCE AND ON ROAD-MAKING. By Major-Gen. W. C. E.
NAPIER. Third Edition. Crown 8vo,~5s.
JEANS (W. T.)—
CREATORS OF THE AGE OF STEEL. Memoirs of
Sir W. Siemens, Sir H. Bessemer, Sir J. Whitworth, Sir J. Brown, and other
Inventors. Second Edition. Crown 8vo, 7s. 6d.
JOKAI (MAURICE)—
PRETTY MICHAL. Translated by R. NISBET BAIN.
Crown 8vo, 5s.
JONES (CAPTAIN DOUGLAS), R.A.—
NOTES ON MILITARY LAW. Crown 8vo, 4s.

JONES. HANDBOOK OF THE JONES COLLECTION
IN THE SOUTH KENSINGTON MUSEUM. With Portrait and Woodcuts. Large crown 8vo, 2s. 6d.

JOPLING (LOUISE)—

HINTS TO AMATEURS. A Handbook on Art. With
Diagrams. Crown 8vo, 1s. 6d.

JUNKER (DR. WM.)—

TRAVELS IN AFRICA. Translated from the German
by Professor KEANE.
Vol. I. DURING the YEARS 1875 to 1878. Containing 38 Full-page Plates
and 125 Illustrations in the Text and Map. Demy 8vo, 21s.
Vol. II. DURING the YEARS 1879 to 1883. Containing numerous Full-page
Plates, and Illustrations in the Text and Map. Demy 8vo, 21s.
Vol. III. COMPLETING THE WORK.   [*In the Press.*

KENNARD (EDWARD)—

NORWEGIAN SKETCHES: FISHING IN STRANGE
WATERS. Illustrated with 30 beautiful Sketches. Second Edition. Oblong
folio, 21s.
Smaller Edition. 14s.

KING (LIEUT.-COL. COOPER)—

GEORGE WASHINGTON. Large crown 8vo. [*In the Press.*

KLACZKO (M. JULIAN)—

TWO CHANCELLORS: PRINCE GORTCHAKOF AND
PRINCE BISMARCK. Translated by MRS. TAIT. New and cheaper Edition, 6s.

LACORDAIRE (PÈRE)—

JESUS CHRIST; GOD; AND GOD AND MAN.
Conferences delivered at Notre Dame in Paris. New Edition. Crown 8vo, 6s.

LAINÉ (J. M.), R.A.—

ENGLISH COMPOSITION EXERCISES. Crown 8vo,
2s. 6d.

LAING (S.)—

PROBLEMS OF THE FUTURE AND ESSAYS.
Eighth Thousand. Demy 8vo, 3s. 6d.

MODERN SCIENCE AND MODERN THOUGHT.
Twelfth Thousand. Demy 8vo, 3s. 6d.

A MODERN ZOROASTRIAN. Fifth Thousand. Demy
8vo, 3s. 6d.

LAMENNAIS (F.)—

WORDS OF A BELIEVER, and THE PAST AND
FUTURE OF THE PEOPLE. Translated from the French by L. E.
MARTINEAU. With a Memoir of Lamennais. Crown 8vo, 4s.

LANDOR (W. S.)—

LIFE AND WORKS. 8 vols.
VOL. 1. Out of print.
VOL. 2. Out of print.
VOL. 3. CONVERSATIONS OF SOVEREIGNS AND STATESMEN, AND
        FIVE DIALOGUES OF BOCCACCIO AND PETRARCA.
        Demy 8vo, 14s.
VOL. 4. DIALOGUES OF LITERARY MEN. Demy 8vo, 14s.
VOL. 5. DIALOGUES OF LITERARY MEN (*continued*). FAMOUS
        WOMEN. LETTERS OF PERICLES AND ASPASIA. And
        Minor Prose Pieces. Demy 8vo, 14s.
VOL. 6. MISCELLANEOUS CONVERSATIONS. Demy 8vo, 14s.
VOL. 7. GEBIR, ACTS AND SCENES AND HELLENICS. Poems.
        Demy 8vo, 14s.
VOL. 8. MISCELLANEOUS POEMS AND CRITICISMS ON THEO-
        CRITUS, CATULLUS, AND PETRARCH. Demy 8vo, 14s.

LANIN (E. B.)—
RUSSIAN CHARACTERISTICS. Reprinted, with Revisions, from *The Fortnightly Review*. Demy 8vo.

LAVELEYE (ÉMILE DE)—
THE ELEMENTS OF POLITICAL ECONOMY.
Translated by W. POLLARD, B.A., St. John's College, Oxford. Crown 8vo, 6s.

LE CONTE (JOSEPH), Professor of Geology and Natural History in the University of California—
EVOLUTION: ITS NATURE, ITS EVIDENCES,
AND ITS RELATIONS TO RELIGIOUS THOUGHT. A New and Revised Edition. Crown 8vo, 6s.

LEFÈVRE (ANDRÉ)—
PHILOSOPHY, Historical and Critical. Translated, with
an Introduction, by A. W. KEANE, B.A. Large crown 8vo, 3s. 6d.

LE ROUX (H.)—
ACROBATS AND MOUNTEBANKS. With over 200
Illustrations by J. GARNIER. Royal 8vo, 16s.

LESLIE (R. C.)—
OLD SEA WINGS, WAYS, AND WORDS, IN THE
DAYS OF OAK AND HEMP. With 135 Illustrations by the Author. Demy 8vo, 14s.

LIFE ABOARD A BRITISH PRIVATEER IN THE
TIME OF QUEEN ANNE. Being the Journals of Captain Woodes Rogers, Master Mariner. With Notes and Illustrations by ROBERT C. LESLIE. Large crown 8vo, 9s.

A SEA PAINTER'S LOG. With 12 Full-page Illustrations
by the Author. Large crown 8vo, 12s.

LETOURNEAU (DR. CHARLES)—
SOCIOLOGY. Based upon Ethnology. Large crown
8vo, 3s. 6d.

BIOLOGY. With 83 Illustrations. A New Edition.
Demy 8vo, 3s. 6d.

LILLY (W. S.)—
ON SHIBBOLETHS. Demy 8vo, 12s.

ON RIGHT AND WRONG. Second Edition. Demy
8vo, 12s.

A CENTURY OF REVOLUTION. Second Edition.
Demy 8vo, 12s.

CHAPTERS ON EUROPEAN HISTORY. With an
Introductory Dialogue on the Philosophy of History. 2 vols. Demy 8vo, 21s.

ANCIENT RELIGION AND MODERN THOUGHT.
Second Edition. Demy 8vo, 12s.

LITTLE (THE REV. CANON KNOX)—
THE CHILD OF STAFFERTON: A Chapter from a
Family Chronicle. New Edition. Crown 8vo, boards, 1s.; cloth, 1s. 6d.

THE BROKEN VOW. A Story of Here and Hereafter.
New Edition. Crown 8vo, boards, 1s.; cloth, 1s. 6d.

LITTLE (THE REV. H. W.)—
H. M. STANLEY: HIS LIFE, WORKS, AND
EXPLORATIONS. Demy 8vo, 10s. 6d

LLOYD (COLONEL E.M.), R.E.—
VAUBAN, MONTALEMBERT, CARNOT: ENGINEER
STUDIES. With Portraits. Crown 8vo, 5s

LLOYD (W. W.), late 24th Regiment—
ON ACTIVE SERVICE. Printed in Colours. Oblong
4to, 5s.

SKETCHES OF INDIAN LIFE. Printed in Colours.
4to, 6s.

LONG (JAMES)—
DAIRY FARMING. To which is added a Description of
the Chief Continental Systems. With numerous Illustrations. Crown 8vo, 9s.

LOVELL (ARTHUR)—
THE IDEAL OF MAN. Crown 8vo, 3s. 6d.

LOW (WILLIAM)—
TABLE DECORATION. With 19 Full Illustrations.
Demy 8vo, 6s.

McCOAN (J. C.)—
EGYPT UNDER ISMAIL: a Romance of History.
With Portrait and Appendix of Official Documents. Crown 8vo, 7s. 6d.

MALLESON (COL. G. B.), C.S.I.—
PRINCE EUGENE OF SAVOY. With Portrait and
Maps. Large crown 8vo, 6s.

LOUDON. A Sketch of the Military Life of Gideon
Ernest, Freicherr von Loudon. With Portrait and Maps. Large crown 8vo, 4s.

MALLET (ROBERT)—
PRACTICAL MANUAL OF CHEMICAL ASSAYING,
as applied to the Manufacture of Iron. By L. L. DE KONINCK and E. DIETZ.
Edited, with notes, by ROBERT MALLET. Post 8vo, cloth, 6s.

MARCEAU (SERGENT)—
REMINISCENCES OF A REGICIDE. Edited from
the Original MSS. of SERGENT MARCEAU, Member of the Convention, and
Administrator of Police in the French Revolution of 1789. By M. C. M. SIMPSON.
Demy 8vo, with Illustrations and Portraits, 14s.

MASKELL (ALFRED)—
RUSSIAN ART AND ART OBJECTS IN RUSSIA.
A Handbook to the Reproduction of Goldsmiths' Work and other Art Treasures
With Illustrations. Large crown 8vo, 4s. 6d.

MASKELL (WILLIAM)—
IVORIES: ANCIENT AND MEDIÆVAL. With nume-
rous Woodcuts. Large crown 8vo, cloth, 2s. 6d.
HANDBOOK TO THE DYCE AND FORSTER COL-
LECTIONS. With Illustrations. Large crown 8vo, cloth, 2s. 6d.

*MASPÉRO (G.), late Director of Archæology in Egypt, and Member of th
Institute of France—*
LIFE IN ANCIENT EGYPT AND ASSYRIA.
Translated by A. P. MORTON. With 188 Illustrations. Crown 8vo, 5s.

MAUDSLAY (ATHOL)—
HIGHWAYS AND HORSES. With numerous Illustra-
tions. Demy 8vo, 21s.

C

GEORGE MEREDITH'S WORKS.

A New and Uniform Edition.   Crown 8vo, 3s. 6d. each.

Copies of the Six-Shilling Edition are still to be had.

ONE OF OUR CONQUERORS.
DIANA OF THE CROSSWAYS.
EVAN HARRINGTON.
THE ORDEAL OF RICHARD FEVEREL.
THE ADVENTURES OF HARRY RICHMOND.
SANDRA BELLONI.
VITTORIA.
RHODA FLEMING.
BEAUCHAMP'S CAREER.
THE EGOIST.
THE SHAVING OF SHAGPAT; AND FARINA.

MERIVALE (HERMAN CHARLES)—

BINKO'S BLUES. A Tale for Children of all Growths.
Illustrated by EDGAR GIBERNE. Small crown 8vo, 5s.

THE WHITE PILGRIM, and other Poems. Crown 8vo, 9s.

MILLS (JOHN), formerly Assistant to the Solar Physics Committee—

ADVANCED PHYSIOGRAPHY (PHYSIOGRAPHIC
ASTRONOMY). Designed to meet the Requirements of Students preparing for
the Elementary and Advanced Stages of Physiography in the Science and Art
Department Examinations, and as an Introduction to Physical Astronomy.
Crown 8vo, 4s. 6d.

ELEMENTARY PHYSIOGRAPHIC ASTRONOMY.
Crown 8vo. 1s. 6d.

ALTERNATIVE ELEMENTARY PHYSICS. Crown
8vo, 2s. 6d.

MILLS (JOHN) and NORTH (BARKER)—

QUANTITATIVE ANALYSIS (INTRODUCTORY
LESSONS ON). With numerous Woodcuts. Crown 8vo, 1s. 6d.

HANDBOOK OF QUANTITATIVE ANALYSIS. Crown
8vo, 3s. 6d.

MILNERS, THE; OR, THE RIVER DIGGINGS. A Story of
South African Life. Crown 8vo, 6s.

MOLESWORTH (W. NASSAU)—

HISTORY OF ENGLAND FROM THE YEAR 1830
TO THE RESIGNATION OF THE GLADSTONE MINISTRY, 1874.
Twelfth Thousand. 3 vols. Crown 8vo, 18s.

ABRIDGED EDITION. Large crown, 7s. 6d.

MOLTKE (FIELD-MARSHAL COUNT VON)—

POLAND: AN HISTORICAL SKETCH. With Biographical Notice by E. S. BUCHHEIM. Crown 8vo, 1s.

MOOREHEAD (WARREN K.)—

WANNETA, THE SIOUX. With Illustrations from Life.
Large crown 8vo, 6s.

MORLEY (THE RIGHT HON. JOHN), M.P.—
RICHARD COBDEN'S LIFE AND CORRESPON-
DENCE. Crown 8vo, with Portrait, 7s. 6d.
Popular Edition. With Portrait. 4to, sewed, 1s. Cloth, 2s.

MURRAY (ANDREW), F.L.S.—
ECONOMIC ENTOMOLOGY. APTERA. With numerous Illustrations. Large crown 8vo, 3s. 6d.

MURRAY (HENRY)—
A DEPUTY PROVIDENCE. Crown 8vo, 3s. 6d.

NECKER (MADAME)—
THE SALON OF MADAME NECKER. By VICOMTE
D'HAUSSONVILLE. 2 vols. Crown 8vo 18s

NESBITT (ALEXANDER)—
GLASS. With numerous Woodcuts. Large crown 8vo,
cloth, 2s. 6d.

NICOL (DAVID)—
THE POLITICAL LIFE OF OUR TIME. Two vols.
Demy 8vo, 24s.

NORMAN (C. B.)—
TONKIN; OR, FRANCE IN THE FAR EAST. With
Maps. Demy 8vo, 14s.

O'BYRNE (ROBERT), F.R.G.S.—
THE VICTORIES OF THE BRITISH ARMY IN
THE PENINSULA AND THE SOUTH OF FRANCE from 1808 to 1814.
An Epitome of Napier's History of the Peninsular War, and Gurwood's Collection
of the Duke of Wellington's Despatches. Crown 8vo, 5s.

O'GRADY (STANDISH)—
TORYISM AND THE TORY DEMOCRACY. Crown
8vo, 5s.

OLIVER (PROFESSOR D.), F.R.S., &c.—
ILLUSTRATIONS OF THE PRINCIPAL NATURAL
ORDERS OF THE VEGETABLE KINGDOM, PREPARED FOR THE
SCIENCE AND ART DEPARTMENT, SOUTH KENSINGTON. With
109 Plates. Oblong 8vo, plain, 16s.; coloured, £1 6s.

OLIVER (E. E.), Under-Secretary to the Public Works Department, Punjaub—
ACROSS THE BORDER; or, PATHAN AND BILOCH.
With numerous Illustrations by J. L. KIPLING, C.I.E. Demy 8vo, 14s.

PAPUS—
THE TAROT OF THE BOHEMIANS. The most
ancient book in the world. For the exclusive use of the Initiates. An Absolute
Key to Occult Science. Translated by A. P. MORTON. With numerous Illustrations. Large crown 8vo, 7s. 6d.

PATERSON (ARTHUR)—
A PARTNER FROM THE WEST. Crown 8vo, 5s.

PAYTON (E. W.)—
ROUND ABOUT NEW ZEALAND. Being Notes from
a Journal of Three Years' Wandering in the Antipodes. With Twenty original
Illustrations by the Author. Large crown 8vo, 12s.

PERROT (GEORGES) and CHIPIEZ (CHARLES)—
A HISTORY OF ANCIENT ART IN PERSIA.
With 254 Illustrations, and 12 Steel and Coloured Plates. Imperial 8vo, 21s.

A HISTORY OF ANCIENT ART IN PHRYGIA—
LYDIA, AND CARIA—LYCIA. With 280 Illustrations. Imperial 8vo, 15s.

A HISTORY OF ANCIENT ART IN SARDINIA,
JUDÆA, SYRIA, AND ASIA MINOR. With 395 Illustrations. 2 vols. Imperial 8vo, 36s.

A HISTORY OF ANCIENT ART IN PHŒNICIA
AND ITS DEPENDENCIES. With 654 Illustrations. 2 vols. Imperial 8vo, 42s.

A HISTORY OF ART IN CHALDÆA AND ASSYRIA.
With 452 Illustrations. 2 vols. Imperial 8vo, 42s.

A HISTORY OF ART IN ANCIENT EGYPT. With
600 Illustrations. 2 vols. Imperial 8vo, 42s.

PETERBOROUGH (THE EARL OF)—
THE EARL OF PETERBOROUGH AND MON-
MOUTH (Charles Mordaunt): A Memoir. By Colonel FRANK RUSSELL, Royal Dragoons. With Illustrations. 2 vols. demy 8vo. 32s.

PIERCE (GILBERT)—
THE DICKENS DICTIONARY. A Key to the Charac-
ters and Principal Incidents in the Tales of Charles Dickens, with Additions by WILLIAM A. WHEELER. New Edition, uniform with the Crown Edition of Dickens's Works. Large crown 8vo, 5s.

PILLING (WILLIAM)—
LAND TENURE BY REGISTRATION. Second Edition
of "Order from Chaos," Revised and Enlarged. Crown 8vo, 5s.

PITT TAYLOR (FRANK)—
THE CANTERBURY TALES. Selections from the Tales
of GEOFFREY CHAUCER rendered into Modern English. Crown 8vo, 6s.

POLLEN (J. H.)—
GOLD AND SILVER SMITH'S WORK. With nume-
rous Woodcuts. Large crown 8vo, cloth, 2s. 6d.

ANCIENT AND MODERN FURNITURE AND
WOODWORK. With numerous Woodcuts. Large crown 8vo, cloth, 2s. 6d.

POOLE (STANLEY LANE), B.A., M.R.A.S.—
THE ART OF THE SARACENS IN EGYPT. Pub-
lished for the Committee of Council on Education. With 108 Woodcuts. Large crown 8vo, 4s.

POYNTER (E. J.), R.A.—
TEN LECTURES ON ART. Third Edition. Large
crown 8vo, 9s.

PRATT (ROBERT), Headmaster School of Science and Art, Barrow-in-Furness—
SCIOGRAPHY, OR PARALLEL AND RADIAL
PROJECTION OF SHADOWS. Being a Course of Exercises for the use of Students in Architectural and Engineering Drawing, and for Candidates preparing for the Examinations in this subject and in Third Grade Perspective conducted by the Science and Art Department. Oblong quarto, 7s. 6d.

PURCELL (the late THEOBALD A.), Surgeon-Major, A.M.D., and Principal Medical Officer to the Japanese Government)—

A SUBURB OF YEDO. With numerous Illustrations.
Crown 8vo, 2s. 6d.

RADICAL PROGRAMME, THE. From the *Fortnightly Review*, with additions. With a Preface by the RIGHT HON. J. CHAMBERLAIN, M.P. Thirteenth Thousand. Crown 8vo, 2s. 6d.

RAE (W. FRASER)—

AUSTRIAN HEALTH RESORTS THROUGHOUT THE YEAR. A New and Enlarged Edition. Crown 8vo, 5s.

RAMSDEN (LADY GWENDOLEN)—

A BIRTHDAY BOOK. Containing 46 Illustrations from Original Drawings, and numerous other Illustrations. Royal 8vo, 21s.

RANKIN (THOMAS T.), C.E.—

SOLUTIONS TO THE QUESTIONS IN PURE MATHEMATICS (STAGES 1 AND 2) SET AT THE SCIENCE AND ART EXAMINATIONS FROM 1881 TO 1886. Crown 8vo, 2s.

RAPHAEL: his Life, Works, and Times. By EUGENE MUNTZ. Illustrated with about 200 Engravings. A New Edition, revised from the Second French Edition. By W. ARMSTRONG, B.A. Imperial 8vo, 25s.

READE (MRS. R. H.)—

THE GOLDSMITH'S WARD; A Tale of London City in the Fifteenth Century. With 27 Illustrations by W. BOWCHER. Crown 8vo, 6s.

REDGRAVE (GILBERT)—

OUTLINES OF HISTORIC ORNAMENT. Translated from the German. Edited by GILBERT REDGRAVE. With numerous Illustrations. Crown 8vo, 4s.

REDGRAVE (RICHARD), R.A.—

MANUAL OF DESIGN. With Woodcuts. Large crown 8vo, cloth, 2s. 6d.

ELEMENTARY MANUAL OF COLOUR, with a Catechism on Colour. 24mo, cloth, 9d.

REDGRAVE (SAMUEL)—

A DESCRIPTIVE CATALOGUE OF THE HISTORICAL COLLECTION OF WATER-COLOUR PAINTINGS IN THE SOUTH KENSINGTON MUSEUM. With numerous Chromo-lithographs and other Illustrations. Royal 8vo, £1 1s.

REID (T. WEMYSS)—

THE LIFE OF THE RIGHT HON. W. E. FORSTER. With Portraits. Fourth Edition. 2 vols. Demy 8vo, 32s.
FIFTH EDITION, in one volume, with new Portrait. Demy 8vo, 10s. 6d.

RENAN (ERNEST)—

THE FUTURE OF SCIENCE: Ideas of 1848. Demy 8vo, 18s.

HISTORY OF THE PEOPLE OF ISRAEL.
FIRST DIVISION. Till the time of King David. Demy 8vo, 14s.
SECOND DIVISION. From the Reign of David up to the Capture of Samaria. Demy 8vo, 14s.
THIRD DIVISION. From the time of Hezekiah till the Return from Babylon. Demy 8vo, 14s.

RECOLLECTIONS OF MY YOUTH. Translated from the French, and revised by MADAME RENAN. Crown 8vo, 8s.

RIANO (JUAN F.)—
THE INDUSTRIAL ARTS IN SPAIN. With numerous
Woodcuts. Large crown 8vo, cloth, 4s.

RIBTON-TURNER (C. J.)—
A HISTORY OF VAGRANTS AND VAGRANCY AND
BEGGARS AND BEGGING. With Illustrations. Demy 8vo, 21s.

ROBINSON (JAMES F.)—
BRITISH BEE FARMING. Its Profits and Pleasures.
Large crown 8vo, 5s.

ROBINSON (J. C.)—
ITALIAN SCULPTURE OF THE MIDDLE AGES
AND PERIOD OF THE REVIVAL OF ART. With 20 Engravings. Royal
8vo, cloth, 7s. 6d.

ROBSON (GEORGE)—
ELEMENTARY BUILDING CONSTRUCTION. Illus-
trated by a Design for an Entrance Lodge and Gate. 15 Plates. Oblong folio,
sewed, 8s.

ROCK (THE VERY REV. CANON), D.D.—
TEXTILE FABRICS. With numerous Woodcuts. Large
crown 8vo, cloth, 2s. 6d.

ROGERS (CAPTAIN WOODES), Master Mariner—
LIFE ABOARD A BRITISH PRIVATEER IN THE
TIME OF QUEEN ANNE. With Notes and Illustrations by ROBERT C.
LESLIE. Large crown 8vo, 9s.

ROOSE (ROBSON), M.D., F.C.S.—
THE WEAR AND TEAR OF LONDON LIFE.
Second Edition. Crown 8vo, sewed, 1s.

INFECTION AND DISINFECTION. Crown 8vo, sewed, 6d.

ROOSEVELT (BLANCHE)—
ELISABETH OF ROUMANIA: A Study. With Two
Tales from the German of Carmen Sylva, Her Majesty Queen of Roumania.
With Two Portraits and Illustration. Demy 8vo, 12s.

ROLAND (ARTHUR)—
FARMING FOR PLEASURE AND PROFIT. Edited
by WILLIAM ABLETT. 8 vols. Crown 8vo, 5s. each.

DAIRY-FARMING, MANAGEMENT OF COWS, etc.
POULTRY-KEEPING.
TREE-PLANTING, FOR ORNAMENTATION OR PROFIT.
STOCK-KEEPING AND CATTLE-REARING.
DRAINAGE OF LAND, IRRIGATION, MANURES, etc.
ROOT-GROWING, HOPS, etc.
MANAGEMENT OF GRASS LANDS, LAYING DOWN GRASS,
ARTIFICIAL GRASSES, etc.
MARKET GARDENING, HUSBANDRY FOR FARMERS AND
GENERAL CULTIVATORS

ROSS (MRS. JANET)—
EARLY DAYS RECALLED. With Illustrations and
Portrait. Crown 8vo, 5s.

ROSS (RONALD)—
THE DEFORMED TRANSFORMED: A Drama in Five
Acts. Crown 8vo, 3s. 6d.

SCHREINER (OLIVE), (RALPH IRON)—

THE STORY OF AN AFRICAN FARM. Crown 8vo,
1s. ; in cloth, 1s. 6d.
A LIBRARY EDITION, on Superior Paper, and Strongly Bound in Cloth. Crown 8vo, 3s. 6d.

SCHAUERMANN (F. L.)—

WOOD-CARVING IN PRACTICE AND THEORY,
AS APPLIED TO HOME ARTS. With Notes on Designs having special
application to Carved Wood in different styles. Containing 124 Illustrations.
Large crown 8vo, 7s. 6d.

SCIENCE AND ART: a Journal for Teachers and Scholars.
Issued monthly. 3d. See page 38.

SCOTT (JOHN)—

THE REPUBLIC AS A FORM OF GOVERNMENT;
or, The Evolution of Democracy in America. Crown 8vo, 7s. 6d.

SCOTT (LEADER)—

THE RENAISSANCE OF ART IN ITALY: an Illus-
trated Sketch. With upwards of 200 Illustrations. Medium quarto, 18s.

SCOTT-STEVENSON (MRS.)—

ON SUMMER SEAS. Including the Mediterranean, the
Ægean, the Ionian, and the Euxine, and a voyage down the Danube. With a
Map. Demy 8vo, 16s.

OUR HOME IN CYPRUS. With a Map and Illustrations. Third Edition. Demy 8vo, 14s.

OUR RIDE THROUGH ASIA MINOR. With Map.
Demy 8vo, 18s.

SEEMAN (O.)—

THE MYTHOLOGY OF GREECE AND ROME, with
Special Reference to its Use in Art. From the German. Edited by G. H.
BIANCHI. 64 Illustrations. New Edition. Crown 8vo, 5s.

SETON-KARR (H. W.), F.R.G.S., etc.—

BEAR HUNTING IN THE WHITE MOUNTAINS;
or, Alaska and British Columbia Revisited. Illustrated. Large Crown, 4s. 6d.

TEN YEARS' TRAVEL AND SPORT IN FOREIGN
Lands; or, Travels in the Eighties. Second Edition, with additions and Portrait
of Author. Large crown 8vo, 5s.

SHEPHERD (MAJOR), R.E.—

PRAIRIE EXPERIENCES IN HANDLING CATTLE
AND SHEEP. With Illustrations and Map. Demy 8vo, 10s. 6d.

SHIRREFF (EMILY)—

A SHORT SKETCH OF THE LIFE OF FRIEDRICH
FROBEL; a New Edition, including Fröbel's Letters from Dresden and Leipzig
to his Wife, now first Translated into English. Crown 8vo, 2s.

HOME EDUCATION IN RELATION TO THE
KINDERGARTEN. Two Lectures. Crown 8vo, 1s. 6d.

SHORE (ARABELLA)—

DANTE FOR BEGINNERS: a Sketch of the "Divina
Commedia." With Translations, Biographical and Critical Notices, and Illustrations. With Portrait. Crown 8vo, 6s.

SIMKIN (R.)—
LIFE IN THE ARMY : Every-day Incidents in Camp,
Field, and Quarters.   Printed in Colours.   Oblong 4to, 5s.

SIMMONDS (T. L.)—
ANIMAL PRODUCTS : their Preparation, Commercial
Uses and Value.   With numerous Illustrations.   Large crown 8vo, 3s. 6d.

SIMPSON (M. C. M.)—
REMINISCENCES OF A REGICIDE.   Edited from
the Original MSS. of Sergent Marceau, Member of the Convention, and
Administrator of Police in the French Revolution of 1789.   Demy 8vo, with
Illustrations and Portraits, 14s.

SINNETT (A. P.)—
ESOTERIC BUDDHISM.   Annotated and enlarged by
the Author.   Sixth and cheaper Edition.   Crown 8vo, 4s.

KARMA.   A Novel.   New Edition.   Crown 8vo, 3s.

SMITH (MAJOR R. MURDOCK), R.E.—
PERSIAN ART.   With Map and Woodcuts.   Second Edition.
Large crown 8vo, 2s.

STANLEY (H. M.) : HIS LIFE, WORKS, AND EXPLORA-
TIONS.   By the Rev. H. W. LITTLE.   Demy 8vo, 10s. 6d.

STATHAM (H. H.—)
MY THOUGHTS ON MUSIC AND MUSICIANS.
Illustrated with Frontispiece of the Entrance-front of Handel's Opera House and
Musical Examples.   Demy 8vo, 18s.

STODDARD (C. A.)—
ACROSS RUSSIA FROM THE BALTIC TO THE
DANUBE.   With Numerous Illustrations.   Large crown 8vo, 7s. 6d.

STOKES (MARGARET)—
EARLY CHRISTIAN ART IN IRELAND.   With 106
Woodcuts.   Demy 8vo, 7s. 6d.
Cheaper Edition, Crown 8vo, 4s.

STORY (W. W.)—
CASTLE ST. ANGELO.   With Illustrations.   Crown
8vo, 10s. 6d.

SUTCLIFFE (JOHN)—
THE SCULPTOR AND ART STUDENT'S GUIDE
to the Proportions of the Human Form, with Measurements in feet and inches of
Full-Grown Figures of Both Sexes and of Various Ages.   By Dr. G. SCHADOW.
Plates reproduced by J. SUTCLIFFE.   Oblong folio, 31s. 6d.

SUVÓROFF, LIFE OF.   By LIEUT.-COL. SPALDING.   Crown
8vo, 6s.

SWIFT : THE MYSTERY OF HIS LIFE AND LOVE.
By the Rev. JAMES HAY.   Crown 8vo, 6s.

SYMONDS (JOHN ADDINGTON)—
ESSAYS, SPECULATIVE AND SUGGESTIVE.   2 vols.
Crown 8vo, 18s.

TAINE (H. A.)—
NOTES ON ENGLAND.   With Introduction by W.
FRASER RAE.   Eighth Edition.   With Portrait.   Crown 8vo, 5s.

TAIT (J. S.)—
WHO IS THE MAN? A Tale of the Scottish Border.
Crown 8vo, 1s. ; in cloth, 1s. 6d

TANNER (PROFESSOR), F.C.S.—
HOLT CASTLE; or, Threefold Interest in Land. Crown 8vo, 4s. 6d.

JACK'S EDUCATION; OR, HOW HE LEARNT FARMING. Second Edition. Crown 8vo. 3s. 6d.

TAYLOR (EDWARD R.), Head Master of the Birmingham Municipal School of Art—
ELEMENTARY ART TEACHING: An Educational and Technical Guide for Teachers and Learners, including Infant School-work ; The Work of the Standards ; Freehand ; Geometry ; Model Drawing ; Nature Drawing ; Colours ; Light and Shade ; Modelling and Design. With over 600 Diagrams and Illustrations. Imperial 8vo, 10s. 6d.

TEMPLE (SIR RICHARD), BART., M.P., G.C.S.I.—
COSMOPOLITAN ESSAYS. With Maps. Demy 8vo, 16s.

THOMSON (D. C.)—
THE BARBIZON SCHOOL OF PAINTERS: Corot, Rousseau, Diaz, Millet, and Daubigny. With 130 Illustrations, including 36 Full-Page Plates, of which 18 are Etchings. 4to, cloth, 42s.

THRUPP (GEORGE A.) and FARR (WILLIAM)—
COACH TRIMMING. With 60 Illustrations. Crown 8vo, 2s. 6d.

THRUPP (THE REV. H. W.), M.A.—
AN AID TO THE VISITATION OF THOSE DIS-TRESSED IN MIND, BODY, OR ESTATE. Crown 8vo, 3s. 6d.

TOPINARD (DR. PAUL)—
ANTHROPOLOGY. With a Preface by Professor PAUL BROCA. With 49 Illustrations. Demy 8vo, 3s. 6d.

TOVEY (LIEUT.-COL., R.E.)—
MARTIAL LAW AND CUSTOM OF WAR; or, Military Law and Jurisdiction in Troublous Times. Crown 8vo, 6s.

TRAHERNE (MAJOR)—
THE HABITS OF THE SALMON. Crown 8vo, 3s. 6d.

TRAILL (H. D.)—
THE NEW LUCIAN. Being a Series of Dialogues of the Dead. Demy 8vo, 12s.

TROLLOPE (ANTHONY)—
THE CHRONICLES OF BARSETSHIRE. A Uniform Edition, in 8 vols., large crown 8vo, handsomely printed, each vol. containing Frontispiece. 6s. each.

THE WARDEN and BAR-CHESTER TOWERS. 2 vols.	THE SMALL HOUSE AT ALLINGTON. 2 vols.
DR. THORNE.	LAST CHRONICLE OF
FRAMLEY PARSONAGE.	BARSET. 2 vols.

LIFE OF CICERO. 2 vols. 8vo. £1 4s.

TROUP (J. ROSE)—
WITH STANLEY'S REAR COLUMN. With Portraits and Illustrations. Second Edition. Demy 8vo, 16s.

VANDAM (ALBERT D.)—
WE TWO AT MONTE CARLO. Second Edition. Crown 8vo, 1s. ; in cloth, 1s. 6d.